Human Factors in Control Room Design

Human Factors in Control Room Design

A Practical Guide for Project Managers and Senior Engineers

Tex Crampin
Liveware Human Factors Ltd
Goodwood, UK

The Handsome TYPE 45 Daring Class Destroyer
MoD/Crown copyright 2016

Registered Office
John Wiley & Sons Ltd, The Atrium, Southern Gate, Chichester, West Sussex, PO19 8SQ, UK

Editorial Office
The Atrium, Southern Gate, Chichester, West Sussex, PO19 8SQ, UK

For details of our global editorial offices, customer services, and more information about Wiley products visit us at www.wiley.com.

Wiley also publishes its books in a variety of electronic formats and by print-on-demand. Some content that appears in standard print versions of this book may not be available in other formats.

Library of Congress Cataloging-in-Publication Data

Names: Crampin, Tex, 1954– author.
Title: Human factors in control room design :
    a practical guide for project managers and senior engineers / Tex Crampin.
Description: First edition. | Hoboken, NJ : John Wiley & Sons Inc., [2017] |
    Includes bibliographical references and index.
Identifiers: LCCN 2016053394 | ISBN 9781118307991 (cloth) | ISBN 9781118535660
    (epub)
Subjects: LCSH: Military engineering–Handbooks, manuals, etc. | Human
    engineering–Handbooks, manuals, etc. | Control rooms–Design and
    construction.
Classification: LCC UG150 .C73 2017 | DDC 620/.46–dc23 LC record available at
    https://lccn.loc.gov/2016053394

Cover Design: Wiley
Cover Image: pozitivstudija/Gettyimages

Set in 10/12pt Warnock by SPi Global, Pondicherry, India

10 9 8 7 6 5 4 3 2 1

Dedication

To my grandfather Herbert George Crampin, Managing Director of the Crampin Steam Fishing Company Ltd from 1911, whose trawling fleet fed the nation and supported the Royal Navy during two World Wars. H.G. Crampin took over and expanded the company whose origins go back to 4 June 1897, when his uncle William Wesney Crampin commissioned his first ship the Ellen Campbell after a distinguished career going back before 1890 when he was a Skipper. The later trawlers were named after famous cricketers with seven letters. Jardine, Leyland, Hammond, Hendren and Larwood were ordered as sister ships in the 1930s; Bradman, Yardley and Statham followed, with the last launched by Sir Freddie Trueman, the England fast bowler, in the 1960s. Many trawlers were sunk during the two World Wars to the extent that the company never fully recovered after the end of World War II. The Nellie Bruce, under the command of Thomas Bell, was torpedoed off Iceland on 30 October 1916. The steam trawler Yardley represented Grimsby at the Royal Navy Portsmouth Spithead Review in 1953. My dedication is also to H.G. Crampin's uncle and mentor, Captain William Wesney Crampin and the crew of the Grimsby sailing smack Conisbro' Castle who risked their lives rescuing the Norwegian square rigger brig Martin Luther (Master J.C. Hansen) of Drammen, Norway, on 18 March 1890, during a severe storm for which Captain Crampin was awarded a medal for bravery by the Norwegian Royal Family. W.W. Crampin had also rescued the crew of a German trawler and was presented with an ebony box, containing a pair of binoculars, by the Kaiser. Also to Herbert William Crampin, H.G. Crampin's son, who was awarded the OBE by Queen Elizabeth II in 1964 for services to the UK fishing industry prior to its sad decline and the sale of the Crampin Steam Fishing Company to the much larger Ross Group in 1965.

Figure 0.1 *Bradman* launched in 1950 held the Grimsby port record for catches of haddock and halibut. *Source*: Reproduced with permission of Liveware HF Ltd.

Figure 0.2 Rescue of the Norwegian brig *Martin Luther* in 1890 by Captain W.W. Crampin in *Conisbro' Castle*. *Source*: Reproduced with permission of Liveware HF Ltd.

Figure 0.3 Medal awarded to Captain W.W. Crampin by the Norwegian Royal Family. *Source*: Reproduced with permission of Liveware HF Ltd.

Figure 0.4 Plaque below the picture of the 1890 rescue. *Source*: Reproduced with permission of Liveware HF Ltd.

Figure 0.5 The *Statham* arriving brand new from Bremerhaven in 1956, one of the largest and most handsome Grimsby trawlers and the largest ever vessel of the Crampin Steam Fishing Fleet. Built to take the raging seas off Iceland and Greenland, she was uniquely adorned with a *Queen Mary* style funnel. *Source*: Reproduced with permission of Liveware HF Ltd.

Contents

About the Author

Tex Crampin was educated at Sedbergh School and Loughborough University before working at GEC-Marconi on the Nimrod and Merlin sonar systems. Tex then moved to Singer Link-Miles where he set up the Human Factors (HF) Group responsible for military applications of the IMAGE visual system for flight simulation and training needs analysis, flying the Harrier XW-267 with Wing Commander Steve Jennings RAF under Armed Reconnaissance No. 3 and other military aircraft in order to establish precise military user needs for low-level ground attack, air-to-air re-fuelling, carrier deck landing and helicopter operations. He is now a Director of Liveware, having founded the company in 1986 during early collaboration with Cambridge Consultants on a military project. Liveware supports the MOD in all aspects of human factors, notably control room design and marine engineering. Tex lectures to MOD staff on military HF design and from 2000 worked on the HF aspects of the design of key operational compartments for RN warships including TYPE 45, the QEC Class of Aircraft Carriers, TYPE 26 and the MARS Tanker. Liveware's current work now includes HF design in the nuclear industry using RN control room experience. He can be reached at tex@livewarehf.com +(44) 07818-420620.

Preface

The title *Human Factors in Control Room Design* will be referred to in this document as 'the Guide'.

The aim of the Guide is to enable rapid access to HF design information and rules of thumb. A small number of references are provided at the back of the Guide which will point readers to further guidance documentation. The intention was to avoid smothering the reader with references and to try to contain as much practical information as possible in one place.

The HF design examples in this Guide are derived from Liveware's experience and in-house prototyping using their software design tool. The HCIs (Human Computer Interfaces) shown are Liveware's generic designs and do not suggest any indicative final implementation but are based mainly on experience in the design of complex Royal Navy warship control rooms. The human factors principles described are fundamentally generic and, in general, can be applied across other industries such as petrochemical, nuclear, police, fire, ambulance, coastguard, etc.

In all endeavours of Human Factors (HF) design, none is more demanding than the development of HCIs for complex systems. An HCI must accord the attention and diligence commensurate with that of a highly skilled jeweller or carpenter and the years of experience necessary in order to attain the high fidelity of design detail needed.

It has been shown that many significant disasters can be attributed to issues related to a lack of attention to human factors in some way. This can manifest itself in poor training, a hostile environment, sub-optimal equipment or task design, or shortcomings in personnel skills and ability through ineffective operator selection. The cost of accidents is usually far more than the small investment that should have been made in human factors to reduce the risks in the first place.

It has been suggested by many experts that some past catastrophic events could have been avoided had sufficient human factors input been applied early in the design process. Examples worthy of scrutiny include:

1) Apollo 13, 13 April 1970;
2) A320 crash at Habsheim Air Show, 26 June 1988;
3) British Midland Boeing 737-400 Flight 92 Kegworth, 8 Jan 1989;
4) Air France Brazil to Paris Flight 447, 1 June 2009;
5) BP Deepwater Horizon Gulf of Mexico, 20 April 2010;
6) Costa Concordia Italian cruise ship, 13 Jan 2012;
7) Germanwings, 24 Mar 2015.

For further information on the subject, please see references [4, 6 and 7].

1

Introduction to the Guide

1.1   Purpose and Scope

The title *Human Factors in Control Room Design – A Practical Guide for Project Managers and Senior Engineers* will be referred to in this document as 'the Guide'.

The Guide aims to provide easy access to practical and objective Human Factors (HF) data in order to achieve rapid and high fidelity control room design. It contains the rudiments of good HF design practice, based on years of experience by the author, in order to undertake complex control room designs quickly and accurately. This Guide does not replace more detailed and textual HF Guidance such as DefStan 00-250 (Ref 1) and other standards, but it does enable a grasp of the key HF 'rules-of-thumb' in order that a busy project team can get on with the design quickly and hit the ground running within the realistic constraints of a 'design advice needed now' commercial and military working environment.

The scope of the Guide makes it applicable to all but the most specialised control rooms. It does not cover, for example, medical operating theatres or precision engineering manufacturing plants although it could easily be adapted to do so with sufficient Subject Matter Expertise (SME) input. It covers the spatial and Human-computer Interface (HCI) aspects of those rapid reaction control rooms typified by teams of civil or military personnel striving for maximum efficiency in information management, safety and mission situational awareness. Thus it applies to control rooms used by Police, Fire, Ambulance or Coastguard personnel; chemical plants, industrial production plants, refineries, oil rigs, RN warships and submarines, Army and RAF tactical control rooms, tri-service and NATO battle command rooms, air traffic control rooms, etc.

The development and advances in technology have allowed plant and equipment monitoring and control to move away from local control panels. Instead of arrays of dedicated controls and displays, modern control rooms are tending

Human Factors in Control Room Design: A Practical Guide for Project Managers and Senior Engineers, First Edition. Tex Crampin.
© 2017 John Wiley & Sons Ltd. Published 2017 by John Wiley & Sons Ltd.

towards centralised remote control via flat screen multifunction displays, sometimes touch screen. However, some dedicated displays and controls should be retained for safety critical functions. This introduces new problems in that information, notably on overall situational awareness, is not readily available throughout the system. Easy-to-use screen navigation, together with easy-to-interpret screen information, is essential in order to maintain optimum system performance, enhanced safety, user comfort and reduced errors. Further, it is essential to determine what screen real estate (sometimes called glass area) will be required, early in the programme.

2

HF Design Process

2.1 Outline Design Process

Effective control room design starts with a sound HF strategy and process (Figure 2.1).

2.2 Detailed Design Process

The following diagram outlines the key HF design stages, explained in more detail below, leading to the development of a Control Room Operating Philosophy (Figure 2.2).

(Alphanumeric numbering is used above to delineate the key design process stages explained in subsequent sub-sections below in order to provide a simple checklist for designers).

A **State System Mission Objectives**

A1 **Define System Aim** – Define the overall aims of the system in terms of what both the equipment and operators are striving to achieve. For example, to provide control and surveillance of all manufacturing processes in order to achieve maximum productivity, safety and user comfort.

A2 **Scope Boundary of System Capabilities** – Draw the whole human-machine system on one sheet of paper, ready to be broken down into human and machine elements through Task Analysis (see D2 below).

A3 **Identify Design Constraints** – Identify any design constraints that dictate the direction of the design. For example, a need for retained or reduced manning; a need for similar interface design to other plants in order to reduce training overheads, etc.

Human Factors in Control Room Design: A Practical Guide for Project Managers and Senior Engineers, First Edition. Tex Crampin.
© 2017 John Wiley & Sons Ltd. Published 2017 by John Wiley & Sons Ltd.

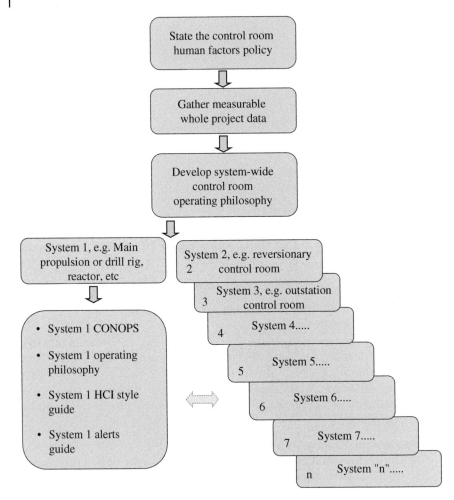

Figure 2.1 Outline HF design process. *Source*: Reproduced with permission of Liveware HF Ltd.

B **Define HF Policy**

    B1 **Define HF Policy and Strategy** – Define the extent of HF involvement and the HF strategy in terms of how HF services will be implemented. For example, ascertain whether there is a policy to reduce manning, introduce touch screen HCIs and reduce workload within one of the highly critical Control Room processes, etc.

    B2 **Identify Relevant HF Standards** – Identify and list the key HF Standards relevant to the design within the industry concerned. For example, DefStan 00-250 and 08-111 for military systems, existing

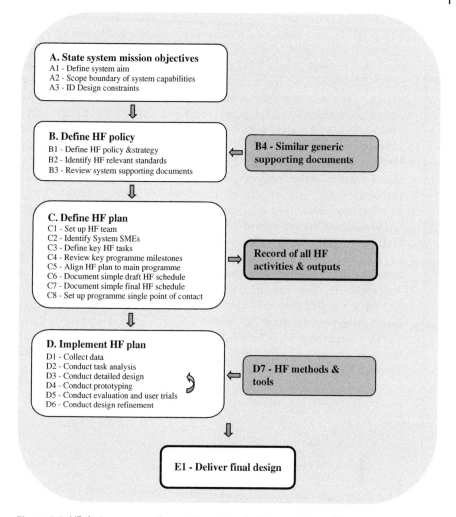

Figure 2.2 HF design process. *Source*: Reproduced with permission of Liveware HF Ltd.

standards in the Oil or Chemical businesses, etc. The purpose of this HF Guide is not to review standards; this must be undertaken as a separate exercise by suitably qualified staff.

B3 **Review System Supporting Documents** – Review any available literature on previous design attempts or retrofits on the specific plant concerned, in order to speed up an understanding of the key HF issues.

B4 **Supporting Documents** – Review any generally available literature on similar control room designs in other industries.

C **Define HF Plan Structure**

C1 **Setup HF Team** – Set up an HF team appropriate to the industry and of a sufficiently small size that is manageable within the total design team. For example, choose HF experts and SMEs who are conversant and experienced with the technologies and HF issues pertinent to the design. Typically, for a large industrial Control Room or Warship design programme, 2–3 HF experts and 1–2 SMEs might be required. The SMEs would most likely be part of the system programme team anyway.

C2 **ID System SMEs** – Select SMEs early in the design process since they are an essential component of design success. Their deployment ensures that systems are designed relevant to the tasks of the users and can bring an enormous amount of experience, thus short-cutting the time taken to 'cut-to-the-chase' in identifying the critical HF issues.

C3 **Define Key HF Tasks** – Analyse only the essential operational tasks since Task Analysis can be very time consuming. These tasks should be agreed with the project team and should include those Mission Critical Tasks (MCTs) that impact system success, are regularly performed and have a high safety and user workload component. The following diagram shows the four key determinants of good system design: **TOSE** – **Tasks** that are achievable by **Operators** chosen and trained for the job, using an optimally designed **System** in a comfortable **Environment** (Figures 2.3, 2.4):

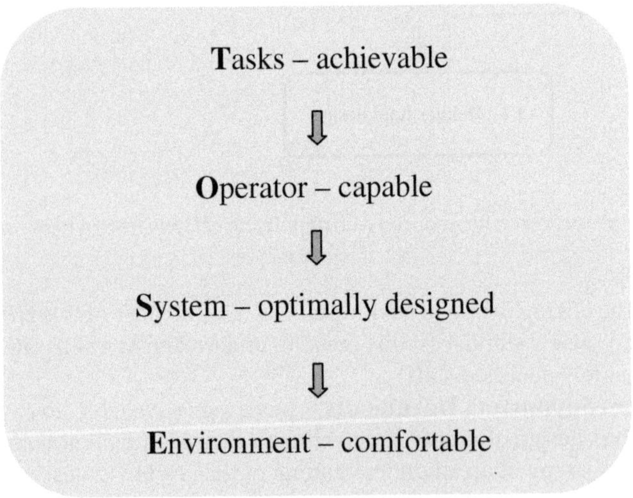

Tasks – achievable

⇩

Operator – capable

⇩

System – optimally designed

⇩

Environment – comfortable

Figure 2.3 Factors impacting human performance. *Source*: Reproduced with permission of Liveware HF Ltd.

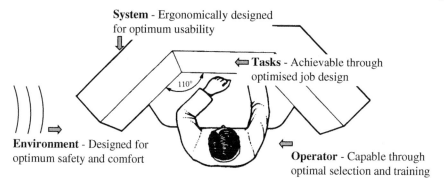

System - Ergonomically designed for optimum usability

Tasks - Achievable through optimised job design

110°

Environment - Designed for optimum safety and comfort

Operator - Capable through optimal selection and training

Figure 2.4 Task - operator - system – environment. *Source*: Reproduced with permission of Liveware HF Ltd.

C4 **Review Key Programme Milestones** – Review the overall control room design schedule then identify where key HF activities should align with that schedule. Discuss this with the project team.

C5 **Align HF Plan to Main Programme** – Ensure that key HF activities are identified early and addressed early in the overall programme schedule. A major factor in system design failure is not getting HF design inputs into the design early enough. Align HF tasks to main programme tasks.

C6 **Document Simple Draft HF Schedule** – Document the aligned HF schedule with the main programme schedule and agree this with the project team.

C7 **Document Simple Final HF Schedule** – Present the completed draft HF Plan to the Project Manager for review, agree key design aspects and document final HF schedule.

C8 **Set up Programme Single Point of Contact** – Choose a single pro-HF member of the overall project team who can represent the HF team fairly at major review meetings, open doors for access to key systems information and project personnel and ensure that HF design issues are thoroughly assessed throughout the programme.

D **Implement HF Plan – Conduct HF Tasks**

D1 **Collect Data**

D1.1 **Identify System Functions** – Identify the main things the system must do, in layman's terms, and create a super-list of Mission Critical Functions (MCFs). For example:

1) Power control and management;
2) Fuel management;
3) Fire alert status;
4) CCTV monitoring;
5) On-line training;
6) Built-in test;etc.

D1.2 **Review Existing HF Inputs** – Obtain access to previous HF design efforts and documentation that may assist in identifying critical HF issues or known problem areas.

D1.3 **Conduct Literature Search** – Review design aspects from parallel industries, scrutinise safety data and identify future display and control technologies that might be relevant to downstream HCI and HMI designs.

D1.4 **Document Data Collection Output** – Document all of the above findings, together with meetings, visits and actions arising, in the HF Plan.

D2 **Conduct Task Analysis**

D2.1 **Breakdown Functions into Activities** – Identify the key functions of the system, for example, the key functions of a smart phone might include phone, contacts, mail, messages, camera, music etc., so it would make sense to have these functions immediately available on a top level page (as the iPhone does). Decompose the MCFs (Mission Critical Functions) down to activity level (Figure 2.5).

Decomposing further than 'Activity' level is likely to generate too much data and should only be done when such detail is vitally necessary. An example of what is meant by the terms Functions, Tasks, etc. is provided below for a High Pressure Salt Water (HPSW) system (Figure 2.6).

An 'Action' is the lowest level of human or machine capability such as operating a switch, detecting a target's presence, making a simple decision or obeying a simple procedure. It is a case of inspecting the breakdown of functions and making a common sense judgement on the degree of decomposition and scrutiny

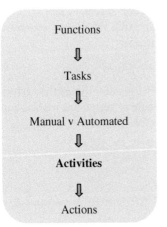

Functions

⇓

Tasks

⇓

Manual v Automated

⇓

Activities

⇓

Actions

Figure 2.5 Breakdown of functions down to actions. *Source*: Reproduced with permission of Liveware HF Ltd.

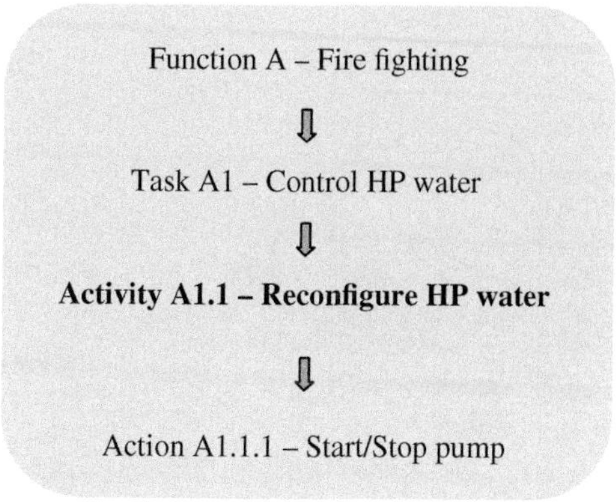

Figure 2.6 Example of functional terms. *Source*: Reproduced with permission of Liveware HF Ltd.

Function A – Fire fighting

⇩

Task A1 – Control HP water

⇩

Activity A1.1 – Reconfigure HP water

⇩

Action A1.1.1 – Start/Stop pump

required in order to effect good design. There are principally four major types of human skill in accomplishing an action:

1) **Psychomotor skill** – muscle-eye coordination, e.g. walking, hitting a ball, pressing a button;
2) **Cognitive skill** – thinking, making a decision, e.g. decide to enter burning compartment;
3) **Procedural skill** – pre-learned response such as carrying out a set of Emergency Operating Procedures e.g. shut down main pump;
4) **Perceptual skill** – sensing, e.g. observing a rise in temperature.

D2.2 **Break down Tasks into Key Mission Critical Activities (MCAs)** – Collate and identify the critical activities which will influence the design approach from the list of activities undertaken by either an individual or team. This will force a filtering of the task analysis data and make the design process shorter, more manageable and more focused on the ultimate design solution.

D2.3 **Conduct User Interviews and Define Manning** – Use the Mission Critical Activities (MCAs) and interviews with key users to drive manning decisions. Typically, in order to save through-life costs, the aim is to reduce manning but ideally, the optimum manning should be calculated in terms of what it takes to complete the activities. The use of simple, reliable basic equipment is sometimes all that is required. Figures 2.7 and 2.8 below show a submarine crew navigating with map and compass, neither of which requires any outside power source and is immune from contamination from computer viruses or network transport delays. Excessive manning reductions can leave systems vulnerable and cost far more in the long run if catastrophic

Figure 2.7 HMS Vanguard on operations with the TYPE 45 Destroyer. *Source*: MoD/Crown copyright 2016.

Figure 2.8 HMS Vanguard: The use of simple, reliable basic equipment is sometimes all that is required. *Source*: MoD/Crown copyright 2016.

Figure 2.9 HMS Albion: Manning needs to take account of different system and operational requirements. *Source*: MoD/Crown copyright 2016.

failures occur; manning needs to take account of different system and operational requirements (Figure 2.9).

D2.4 **Define Roles and Information Requirements** – Define operator roles. Most of these are likely to be readily evident from the current design but will need to be adjusted and refined. Further, for each role, the information requirements must be derived so that the right information is available at the right time, usually visually on electronic screens. However, the information needed could manifest itself in other sensory devices such as alert sounds

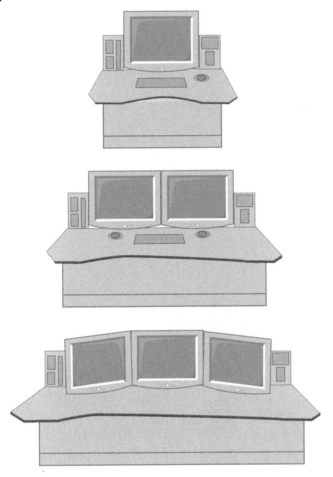

Figure 2.10 Single or multi-screen consoles. *Source*: Reproduced with permission of Liveware HF Ltd.

or vibrations, for example, a stick shaker in a jet to warn of an incipient stall. In most cases, an early key decision is deciding on the screen real estate required for each operator or team and which displays and controls should be available permanently (Figure 2.10).

D2.5 **Develop Mission Critical Scenarios (MCSs) and Scripts –** Identify, or for a new system develop, the Standard and Emergency Operating Procedures (in military terms SOPs and EOPs). These fall out of the key functions of the system and are most likely to lead to a shortlist of key Scenarios made up of the tasks and activities identified above. It is worth spending some time on this since, from experience, they have a long shelf life, usable downstream in assessing the design with prototyping and simulations and in developing definitive SOPs and EOPs for the final design, including training. A scripted scenario is provided below as a simple Royal Navy example in order to follow through the benefits of the task analysis data:

A Scenario is envisaged whereby a Lynx helicopter flying exercise is planned from 09.00 to 11.00 on board a typical TYPE 23 Frigate. Emphasis is on recovery of the helicopter:

07.30 **Brief** - includes key crew members

08.00 **Command** - authorises flight

08.05 **Range** - walking the helicopter out of the hangar

08.10 **Prepare for Flying** - FOD (Foreign Objects and Debris) clearance, re-fuelling

08.40 **Hands to Flying Stations** - FDO checks weather operating limits

08.55 **Alert State AL5** - Lynx at alert

09.00 **Launch**

10.45 **Recovery begins**

10.45 **Command** - informs AC of intention to recover

10.45 **Aircraft Controller** - informs OOW of decision to recover

10.46 **Officer of the Watch** - calculates flying course

10.47 **Officer of the Watch** - pipes 'Hands to Flying Stations' 15 mins to go to recovery

10.50 **Flight Deck Officer** - assembles Flight Deck team

10.51 **Flight Deck Officer** - checks personnel and Flight Deck condition

10.53 **Flight Deck Officer** - calls OOW 'Bridge – Flight Deck is ready for flying'

10.55 **Flight Deck Officer** - calls OOW when Lynx is downwind for permission to land

10.56 **Officer of the Watch** - gives Green light

10.56 **Flight Deck Officer** - calls OOW 'Roger Green'

10.57 **Aircraft Controller** - controls Lynx to a gate 2nm from ship within a 30^0 cone on Red 160^0

10.58 **Lynx** - reduces speed to 60 kts

10.59 **Flight Deck Officer** - takes over control of recovery at 0.25nm/125 feet

10.59 **Aircraft Controller** - calls 'FDO Control'

11.00 **Flight Deck Officer** - steps forward to marshal helicopter to Flight Deck

D2.6 **Conduct Spatial Link Analysis and Organisational Hierarchy** – Establish the spatial relationships between individual operators and teams of operators. This is achieved by inspection, SME interviews, tabulation and link analysis of where tasks are shared from the above task analysis. The following diagrams provide an example table of operator tasks and ensuing spatial link analysis. This graphical representation makes it easier to formulate a workable early system design, in conjunction with user Operability Assessments. Table 2.1 judges the interaction between operators in terms of the degree of:

1) **Frequency (F)** - how often;
2) **Duration (D)** - for how long;
3) **Importance (I)** - mission criticality.

This method works for any scope of system from a small number of users in a small control room to large numbers of users in several control rooms, spatially separated. The example below (Table 2.1, Figure 2.11) is taken from a real military design exercise whereby the brief was to design a visual display for showing ship pitch and roll in order to judge whether a safe helicopter landing is possible. First, it was necessary to understand the 'system' for landing helicopters safely onto the flight deck of a Royal Navy frigate in terms of crew tasks. The key team of crew members comprised:

1) **CO** - Command;
2) **OOW** - Officer of the Watch;
3) **AC** - Aircraft Controller;
4) **CFD** - Captain of the Flight Deck;

Table 2.1 Example crew spatial link table.

	CO			OOW			AC			CFD			FDO			FDC			Pilot		
	F	D	I	F	D	I	F	D	I	F	D	I	F	D	I	F	D	I	F	D	I
CO				–	–	•	–	–	•	–	–	–	–	–	–	–	–	–	–	–	–
OOW	–	–	•				–	–	•	–	–	–	•	–	–	–	–	–	–	–	–
AC	–	–	•	–	–	•				–	–	–	–	–	•	–	–	–	•	•	•
CFD	–	–	–	–	–	–	–	–	–				•	•	•	•	•	•	–	–	–
FDO	–	–	–	•	–	–	–	–	•	•	•	•				–	–	•	–	–	•
FDC	–	–	–	–	–	–	–	–	–	•	•	–	–	–	•				–	–	–
Pilot	–	–	–	–	–	–	•	•	•	–	–	–	–	•	•	–	–	–			

Source: Reproduced with permission of Liveware HF Ltd.

F: Frequent interaction D: Regular duration of interaction I: High importance interaction

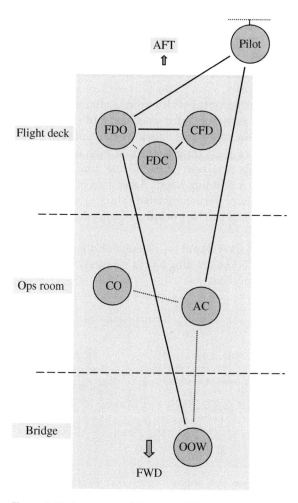

Figure 2.11 Example spatial operator link analysis. *Source*: Reproduced with permission of Liveware HF Ltd.

5) **FDO** - Flight Deck Officer;
6) **FDC** - Flight Deck Crew;
7) **Pilot** - Pilot of Lynx Helicopter.

D2.7 **Scrutinise High Performance Activities** – Focus on the high performance Mission Critical Activities (MCAs) otherwise, HF analyses can be overly time consuming. These are activities which have the most impact on mission success or control room effectiveness, for example, activities that are safety oriented, waste valuable operator time or hold up production processes incurring downstream costs. Assess tasks that are also performed regularly.

D2.8 **Assess Workload of MCAs** – Scrutinise the critical activities in detail; they are worth valuable HF attention to get right. This scrutiny may require an assessment of operator workload, usually involving both human and machine activities. Any data collected will have a long shelf life, usually years, since it can be re-visited downstream for retrofits or re-design purposes with minimal alteration. There are many techniques to assess operator workload and the best are the simplest, focusing on face value, obvious bottlenecks, over-complicated tasks or, most likely, poorly designed equipment.

D2.9 **Conduct Training Needs Analysis** – Use the task analysis data to establish training needs. A Training Needs Analysis (TNA) requires a Task Analysis (TA) to have been completed. The key processes of undertaking a TNA are summarised below in Figure 2.12 and discussed in detail in Chapter 9.

D2.10 **Document Task Analysis Output** – Document the TA (Task Analysis), including the TNA (Training Needs Analysis), in text and graphical format in the HF Plan.

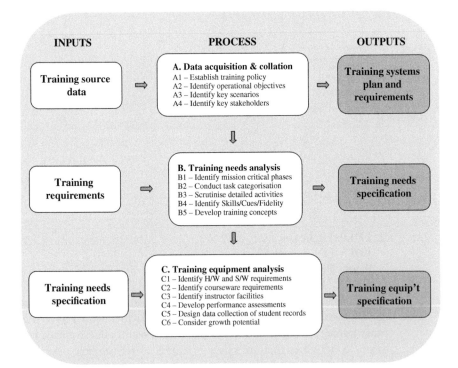

Figure 2.12 Example TNA process. *Source*: Reproduced with permission of Liveware HF Ltd.

D3 **Conduct Detailed Design**

D3.1 **Reallocate Functions** – Section D2.1 above discussed the process of identifying and breaking down MCFs (Mission Critical Functions) into Tasks and Activities. As a result of the TA, it is likely to have transpired that new functions have been identified or functions that were thought to have been trivial are now deemed critical. This is the reason for function reallocation.

D3.2 **Establish Outline Control Room Operating Philosophy** – At this stage of the design process a comprehensive understanding of both the existing system and the intended new control room design will have been derived from function and task analyses. It should therefore be possible to begin to map out how the control room will be operated in terms of a Control Room Operating Philosophy. This should be done as early as possible because it provides a design to be 'kicked around' and discussed as a 'straw model' by key Stakeholders. The Control Room Operating Philosophy should provide, for example, the outline manning (Section D2.3), the key functions performed (Section D3.1) and approximate screen real estate for the control room should be known (Section D2.4). Then begin to define the concept of operation for each operator and therefore scope the control room mode of operation. The initial Control Room Operating Philosophy is likely to manifest itself in terms of a system architecture diagram (Figure 2.13), control room staff hierarchy usually comprising Director, Supervisory and Operator levels of information flow and an outline control room spatial layout (Figures 2.14, 2.15).

D3.3 **Establish Outline Control Room HCI Operating Philosophy** – Design top level pages showing access to all MCFs (Mission Critical Functions), either as a simple menu list, graphical group of different and unrelated functions exemplified by the excellent iPhone HCI (Figure 2.16) or, where spatial appreciation is required in a major industrial plant, a top level system mimic (Figure 2.17). A PDA (Personal Digital Assistant) HCI is less of a spatial device than a control room HCI and therefore requires a different top level page hierarchy. This process will consolidate the earlier estimates of screen real estate for each operator and therefore the whole control room.

D3.4 **Design Outline Control Room Layout** – This is a spatial exercise based on a knowledge of the outline manning, operator roles, key functions and activities, critical scenarios and link analysis showing the different operator spatial relationships. Start with the link analysis which will help to establish priority operator positions, groups of functionally related operators, eye-balling and shared screen requirements. A more detailed example of this design process is provided in Chapter 3.

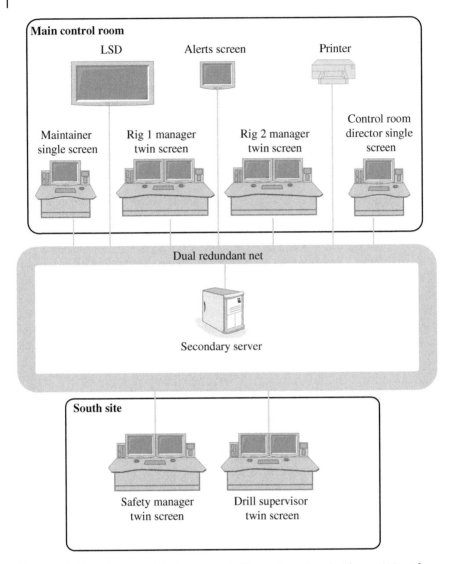

Figure 2.13 Control room architecture example. *Source*: Reproduced with permission of Liveware HF Ltd.

D3.5 **Design Outline Console Configurations** – First establish how many different types of console are required, e.g. single or multi-screen, seated, standing, shared screen information, shared eye-balling, etc. Different configurations suit different control room tasks and information requirements. Whilst console commonality is an obvious objective in terms of short term cost saving, a common

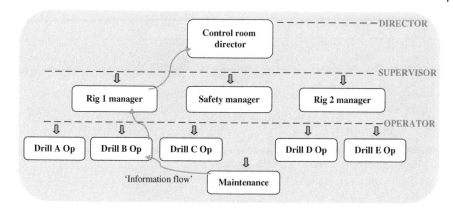

Figure 2.14 Control room staff hierarchy example. *Source*: Reproduced with permission of Liveware HF Ltd.

Figure 2.15 The bridge of ice patrol vessel HMS Protector is pictured lit up at night during a transit in the Antarctic. *Source*: MoD/Crown copyright 2016.

trap is trying to use a 'one-size-fits-all' solution that can severely inhibit operational safety and success.

D3.6 **Design Outline Consoles** – From the candidate outline console configurations, select the configuration(s) that best meets the functional needs of the control room, eg, single or multi-screen, seated,

Figure 2.16 Example simple graphics top level HCI. *Source*: Reproduced with permission of Apple. Inc.

standing, shared screen information, shared eye-balling, etc. The rules for console design are very similar, irrespective of the configuration, but some workstations are likely to need special features to suit the tasks of its host operator, for example, propulsion levers in a ship machinery control room, reactor control levers in a nuclear power station control room, the main helm on a ship, the flying controls on a jet, etc.

D3.7 **Design Outline Panels** – Single operator console panel layout design should rely heavily upon the information generated from task analysis. This will affect how the panels are structured in terms of display/control relationships, grouping, sequences, etc. The three

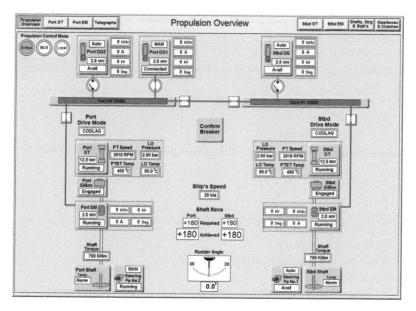

Figure 2.17 Example graphics top level HCI mimic. *Source*: Reproduced with permission of Liveware HF Ltd.

key criteria used when siting controls and displays are frequency of use, duration of use and importance of the devices.

D3.8 **Design Outline Equipment** – Equipment refers to those items mounted on or near the operator consoles which are important but indirectly related to the functioning of the main control room systems. Such equipment might include fire extinguisher mountings and locations, coat hangers, cup holders, fuse boxes, storage cabinets, supervisory and training staff seating, light switches, etc.

D3.9 **Design Outline HMIs (Displays and Controls)** – Build on the outline panel design above to create outline console HMIs showing the general disposition of displays and controls relative to the seated operator.

D3.10 **Design Outline HCIs** – HCIs should be designed following data derived from Task Analysis. Task Analysis provides assistance to where design precision should be applied. In broad terms HCI design should cover:

1) HF design requirements;
2) System overview;
3) Menus;
4) Windows;
5) Controls;

6) Dialogue;
7) Text boxes;
8) Checklists;
9) Tabs;
10) Text;
11) Terminology and abbreviations;
12) Symbology and icons;
13) Pointers and cursors;
14) Colour;
15) Control panel pages;
16) Mimics;
17) Alerts;
18) Auditory displays;
19) Station-in-control;
20) Modes of operation;
21) Emergency and reversionary operation.

D4 **Conduct Prototyping**

D4.1 **Choose an Appropriate Prototyping Tool** – prototyping tools must be:

1) Easy to use, preferably by an HF expert whom would be able to utilise the tool to the best HF effect;
2) Able to undertake rapid changes to a design, in front of subject matter experts, at the time of review in order to achieve rapid reaction to design feedback. For example, if a user decides that a light background mimic shade is best then this should be able to be demonstrated and tested there and then, not hours later;
3) Of sufficient fidelity to represent the simulated tasks that a real operator would generally encounter and execute on the real system. For example, if screens appear within 200 ms on the real system then they should on the prototype; if alerts flash at 5 Hz on the real system then they should on the prototype, and so on;
4) Of sufficient fidelity to have face validity with the user and therefore credibility as an evaluation design tool.

D4.2 **Identify Critical Design Issues** – Identify those aspects of the HMI/HCI that require focused prototyping to challenge or test the layout and concept of operation. Such aspects might include:

1) Those where problems have occurred in past designs to the detriment of safety, performance or comfort; for example, slow reaction to emergency scenarios, slow conduct of operation with high error rates, poor seating, etc.;
2) Those that represent a new method of operating the systems and therefore deviate from previous user experience;

3) Those that represent the fundamental core of system operation, where prototyping of one function can be used as a test for several other functions and therefore achieve efficiency within the design tool.

D4.3 **Document Prototyping Output** – The best documented prototyping results are based around a comprehensive representation of the real system, that is, an actual working prototype itself. Documented comments can thus be written around the prototype to explain why it is designed the way it is.

D5 **Conduct Evaluation and User Trials**

D5.1 **Design User Trials** – The starting point for user trials is a set of detailed scripted scenarios representing all aspects of operating the systems in the control room. For example, a scenario might include testing the following:

1) Console log on and log off;
2) Fire in a Diesel Generator (DG) enclosure module requiring:
   a) Isolation of systems and assessment of the impact on other services;
   b) Execution of 'safe-to-work' procedures;
   c) Transfer to an alternative console when user's own console fails.
3) Evolution exercising security protection operations;
4) Run-through of a precisely scripted scenario (Table 2.2) previously tabulated with the assistance of SMEs:

Table 2.2 Example scripted scenario: Fire in diesel generator.

Ref	Events/Activities
01	A fire danger occurs on the DG (Diesel Generator) alternator. Fire danger is as a result of the lubricating cross connection pipe failing and spraying oil onto a hot exhaust
02	System displays DG Lub Oil Alert
03	Operator 1 - accepts DG lub oil Alert
04	Operator 2 - transmits emergency running conditions to supervisor
05	Operator 2 - selects reversionary control using hard wired buttons on console
06	Operator 3 - starts auxiliary DG
07	Operator 3 - briefs supervisor via comms
08	Operator 3 - sends out runner to investigate
09	Etc.

Source: Reproduced with permission of Liveware IIF Ltd.

D5.2 **Evaluate Prototypes with Users** – This should be achieved with planned user trials whereby experienced operators are guided through a complete set of scenarios in order to evaluate how each console HMI/HCI responds. This is undertaken for each individual console whereupon teams of operators can be brought in to enact sub-team and then complete control room exercises using the previously derived scripted scenarios. Full control room evaluation is achieved by conducting operability assessments which are described in the next section.

D5.3 **Conduct Operability Assessments** – Full control room operability trials should be assessed with a full-size mock-up. Depending on the stage of control room development this can be achieved at varying levels of fidelity. Initially this can be a paper exercise with hard copy screen information available in front of each operator such that pages are turned to emulate selecting a new screen electronically. It is recommended that this initial paper exercise is conducted with mock-up consoles (they could just be desks) spatially arranged as close to the intended layout as possible. In this way, a more accurate evaluation will emerge of how suitable the straw model layout is, and how easily the operators can interact and eye-ball each other. Following paper trials, a better control room layout can be derived and the trials re-run using actual working prototypes of the HCIs and HMIs.

Whatever the level of trials fidelity, a common approach should be adopted. Operability assessments should be planned and executed to a high degree of precision with each stakeholder being provided with an information pack containing:

1) **Executive Summary** – outlining the purpose of the trials and contents of the Information Pack;
2) **Schedule** – description of the events usually spanning a two-day evaluation;
3) **Process Description** – explaining the aim and HF issues to be evaluated;
4) **Spatial Layout** – diagram of the control room together with descriptions of each operator's position, role, interactions and key functions;
5) **Scenarios** – tables of a suite of scenarios (about a dozen is typical) showing the events unfolding along a timeline;
6) **Data Collection Forms** – simple one-page forms, to be filled in by each stakeholder, asking specific questions about the control room design;
7) **Project Schedule** – a single sheet schedule showing each stakeholder where the team has got to in terms of control room design development;

8) **Checklist** – a simple list of everything required to undertake the operability assessment and make the event a success;

9) **Annex** – containing records of previous design evaluation exercises, meeting notes and key design decisions, etc.

D5.4 **Document User Trials Output** – these are best documented in a separate trials report and lodged within the HF Plan.

D6 **Conduct Design Refinement**

Following detailed design completion, all aspects of the design solution can be refined and a comprehensive HCI Style Guide produced, that is, describing the precise look and feel of the user interface:

D6.1 **Define Control Room Operating Philosophy** – The Control Room Operating Philosophy is a high level concept outlining how systems should be operated ('driven'). For example, one concept might be that the control room is operated by two seated operators during quiet, non-critical states each using three-screen consoles. The centre screen might provide an overview of plant status, the right hand screen a list of alerts and the left hand screen available for drilling down into plant detail when required. This is one of many outline operating philosophy concepts. From this, a detailed operating philosophy can be derived.

D6.2 **Establish Detailed Operating Philosophy** – The following list summarises the key factors required in developing a detailed Operating Philosophy:

1) **Functional Software Components** – Most complex control rooms are built up of key functional software components. For example, in a Warship PMS (Platform Management System) these might include Machinery Control and Surveillance, Damage Control and Administration. Some control rooms may have many more key software functions but to the user it is essential that these components, however many, are perceived as an integrated package, the components of which all obey the same basic operating rules with similar look and feel. This inspires consistency, confidence and predictability, thus enhancing performance, comfort and safety;

2) **Single Screen Navigation and Control** – The operating philosophy for a single screen should be consistent with that of the left hand screen of a twin screen philosophy. The following design features should be applied:

a) The Primary Navigation bar is positioned on the right hand side of the screen;

b) The single pointer on the screen is controllable by a trackball;

c) Control of equipments is permissible on any of the display pages if the system allows control of that equipment and the operator has Station-in-Control.

Two single screen workstations can be sited side by side and used as a twin screen workstation, but a single operator would have to use two different trackballs to control the two cursors (or a touch screen if appropriate). Operating philosophies need developing for other aspects including:

a) Twin screen navigation and control;
b) Large screen displays;
c) Page hierarchy;
d) Alerts philosophy;
e) Plant start-up and shut-down;
f) Shift/watch changeover;
g) CCTVs;
h) BITE;
i) Training;
j) Data recording, logging and trends;
k) Email;
l) Interlocks and overrides;
m) Administration;
n) Help;
o) Modes.

D6.3 **Refine Control Room Layout** – Using scenarios developed with SME support, test the layout and make any necessary alterations.

D6.4 **Refine Console Configurations** – Using data from task analyses with SME support, test the console layouts and make any necessary alterations.

D6.5 **Refine Console Design** – Using data from sound ergonomics principles, notably anthropometrics and workstation layout, test the console design and make any necessary alterations.

D6.6 **Refine Panel Design** – Using data from sound ergonomics principles, notably control and display layout, test the panel design and make any necessary alterations.

D6.7 **Define Precise Equipment Locations** – Recognise the importance of small and sundry items that can impact operator performance, notably cup holders, desk space, colour, aesthetics, etc.

D6.8 **Design Detailed HMIs (Displays and Controls)** – Refine the concepts derived for HMIs using the information in Chapter 4.

D6.9 **Design Detailed HCIs** – Refine the concepts derived for HCIs using the information in Chapter 5.

D6.10 **Write Operating Philosophy and HCI Style Guide** – An Operating Philosophy starts with a CONOPS (Concept of Operations) which provides an outline description of how operators will use a system in responding to various events. Notably, it will describe how operators navigate through the system, either by electronic pages or by some other means. A CONOPS should detail who does what, when and where and be tested through prototyping using operational scenarios. The CONOPS can quickly be expanded into a full system operating philosophy which is described in Chapter 5. From the Operating Philosophy an HCI Style Guide can be created which provides detailed screen designs and is also described in Chapter 5.

D6.11 **Document Design Output** – The complete design must be documented both electronically and in hard copy to suit different design styles of Suppliers (companies responsible for delivering control room design solutions and products). Most importantly, the documentation must be liberal in the use of graphics and design diagrams which users of the intended system will understand. Avoid dull, lengthy tables of requirements and rules that will not make interesting reading.

D7 **HF Methods and Tools** – There are various HF methods and tools available, some more usable than others. An example, using some of the more practical methods and tools, is provided above in Section D1. Key HF methods and tools include:

1) **Function Analysis** – involves starting the design process by identifying the major system functions;

2) **Task Analysis** –involves expanding the design process by identifying the major tasks attributed to each function; these could be either manual or automatic;

3) **Scenario Analysis** – involves identifying the key scenarios within which the system must operate. These will be very useful later on for assessment and testing;

4) **User Interviews** – involves exploring and identifying the key operator tasks which can then be broken down into activities and, in some cases, actions;

5) **Role and Manning Assessment** – involves identifying the major operator roles and how each compartment and workplace is manned. It would be important here to refer back to any policies on manning, eg, a need to reduce manning;

6) **Link Analysis** – involves identifying the key links between operators to assist in the compartment layout in terms of the geography of the seating, eye-balling, information sharing, etc.;

7) **Workload Assessment** – involves identifying those important and difficult tasks that require more scrutiny, in which case a breakdown to action level may be required;

8) **Training Needs Analysis (TNA)** – involves using the task analysis to form the basis for a TNA (Training Needs Analysis): see Chapter 7.

E **Deliver Final Design and Record in HF Plan**
   E1 **Deliver Final Design**
        E1.1 **Present Design to Project Managers;**
        E1.2 **Review and Implement Changes;**
        E1.3 **Document Final Design Output.**
   E2 **Record in HF Plan.**

3

Workspace Human Factors

3.1 Outline Design Approach

This chapter is about the design of the general space within which a control room team would typically work. On a warship it would be one of the key operational compartments, on an oil rig the main control room and for emergency services an accident or traffic control centre, for example.

The scope of this chapter covers:

1) Workspace design and traffic flow;
2) Workplace design and console configuration;
3) Workstation, console design and anthropometrics;
4) Seating;
5) Maintenance;
6) Co-location.

3.2 Workspace Design and Traffic Flow

3.2.1 Design Outline Control Room Layout

This is a spatial exercise based on a knowledge of the outline manning, operator roles, key functions and activities, critical scenarios and link analyses showing the different operator spatial relationships. Start with the link analysis which will help to establish priority operator positions, groups of functionally related operators, eye-balling and shared screen requirements. Example outline control room layouts are shown in Figures 3.1 and 3.2. They show a warship SCC (Ship Control Centre) fully manned at action and minimally manned at cruise, with the following specific features pertinent to the action state:

1) Space required, including traffic flow, for this number of operators using this size of equipment in this amount of space;

Human Factors in Control Room Design: A Practical Guide for Project Managers and Senior Engineers, First Edition. Tex Crampin.
© 2017 John Wiley & Sons Ltd. Published 2017 by John Wiley & Sons Ltd.

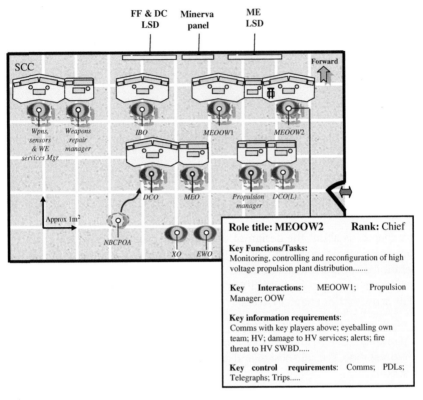

Figure 3.1 Outline control room layout fully manned. *Source*: Reproduced with permission of Liveware HF Ltd.

2) Key operator seated/standing positions with an example role description explaining key functions and tasks, interactions, information and control requirements. This role information is essential for several reasons:

a) It provides a check that every required operator has been accounted for and is aligned to the complement model for that organisation;

b) It provides a summary understanding to all Stakeholders of what each operator's function is in the control room; this is useful information where reduced manning is a central policy;

c) It forms the basis for a TNA (Training Needs Analysis) which would require an understanding of operator roles and tasks;

d) It has a shelf life of many years because this information is useful to many Stakeholders at different points in a control room's design history from development through assessment to acceptance, in-service use, retrofits and disposal.

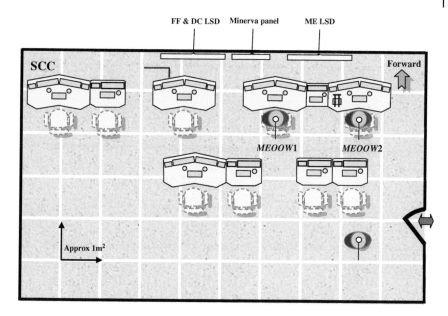

FF & DC LSD   Minerva panel   ME LSD

SCC

Forward

MEOOW1   MEOOW2

Approx 1m²

Figure 3.2 Outline control room layout minimally manned. *Source*: Reproduced with permission of Liveware HF Ltd.

3.2.2 Design Outside Gantry Access and Layout

The following Figures 3.3–3.6 provide some example dimensions for general access.

3.2.3 Optimise Spatial Dimensions to Promote Good Traffic Flow

Ensure operator access and egress by showing seated operators and the space required for seat swivel/pushback and access between (Figure 3.7).

3.2.4 Design Outline Equipment

Equipment refers to those items mounted on or near the operator consoles which are important but indirectly related to the functioning of the main control room systems. Such equipment might include fire extinguishers and their mountings, coat hangers, cup holders, fuse boxes, storage cabinets, supervisory and training staff seating, light switches, etc. The following broad rules should be used when siting control room equipment:

1) Use basic ergonomics principles for reach, clearance, access etc. within the design population, usually 5% females to 95% males;
2) Refer to the DefStan 00-250 (Ref 1) for general placement of items;

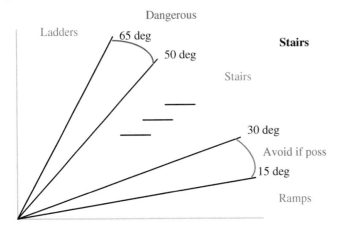

Figure 3.3 Ramps and stairs. *Source*: Reproduced with permission of Liveware HF Ltd.

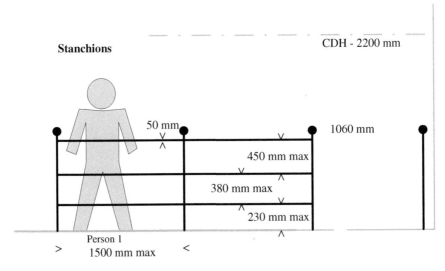

Figure 3.4 Railings 1060 mm height advised (CDH: Clear Deck Height). *Source*: Reproduced with permission of Liveware HF Ltd.

3) Use task analysis data to ensure that the use of periphery equipment is integrated into the use of the primary equipment;

4) Fire extinguishers should be next to exits but raised off the floor and not in the way as a tripping or snagging hazard; raising items above the floor also makes cleaning quicker.

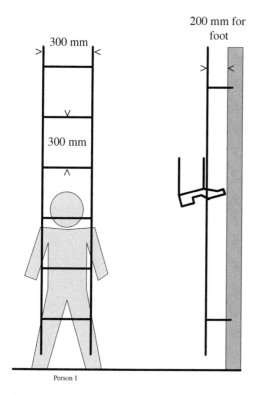

300 mm

200 mm for foot

300 mm

Person 1

Figure 3.5 Ladders. *Source*: Reproduced with permission of Liveware HF Ltd.

Fire fighting zone envelope

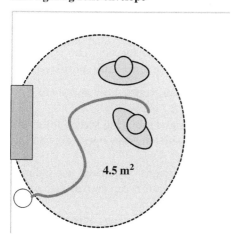

4.5 m²

Figure 3.6 Fire fighting team zone envelope. *Source*: Reproduced with permission of Liveware HF Ltd.

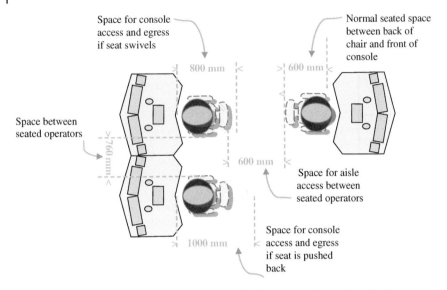

Space for console access and egress if seat swivels

Normal seated space between back of chair and front of console

800 mm

600 mm

Space between seated operators

760 mm

600 mm

Space for aisle access between seated operators

Space for console access and egress if seat is pushed back

1000 mm

600 mm

Figure 3.7 Key spatial dimensions to enhance traffic flow. *Source*: Reproduced with permission of Liveware HF Ltd.

3.3 Workspace Design and Console Configuration

3.3.1 Design Outline Console Configurations

First establish how many different types of console are required, e.g. single or multi-screen, seated, standing, shared screen information, shared eye-balling, etc. Different configurations suit different control room tasks and information requirements. Whilst console commonality is an obvious objective in terms of short term cost saving, a common trap is trying to use a 'one-size-fits-all' solution that can severely inhibit operational safety and success. The following diagrams show some example configurations with their relative merits:

1) Establish basic spatial dimensions for consoles, traffic flow and seating to promote operator access and egress (see Fig 3.7 above).
2) Establish multi-operator seated console designs requiring screen information sharing, depicted by the dotted arrows showing how operators can see the adjacent screen information (Figure 3.8).
3) Establish multi-operator seated console designs requiring eye-balling, depicted by the dotted arrows showing how operators can eyeball each other (Figure 3.9).
4) The Supplier should provide detailed scale layout drawings of the compartments and spaces with system workstations and equipments. The layout drawings should ensure that all equipments fitted into the compartment are

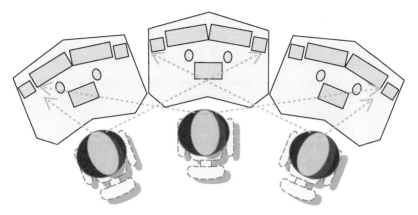

Figure 3.8 Shared screens multiple operator seated console. *Source*: Reproduced with permission of Liveware HF Ltd.

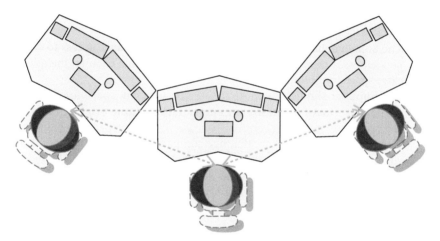

Figure 3.9 Shared eye-balling multiple operator seated console. *Source*: Reproduced with permission of Liveware HF Ltd.

detailed and that there is space within the compartment for Command huddles, workstation maintenance, cabinets, etc.

5) The following compartment HF Operability Statements are mandatory for Suppliers to implement:

   a) Where safety and practicality permits, operator positions should adopt a forward-facing philosophy, that is, facing the bow of a vessel, with controls and displays in the correct fore/aft orientation. For a land-based plant, choose an orientation that best supports an operator's mental model of the outlying plant site. It may be possible to provide a view of the site through windows or CCTV. Beware of sunlight and the fact that, in the northern hemisphere, the sun tracks through south so that a north facing visual control room avoids glare, e.g. an airport VCR (Visual Control Room);

b) Whatever the operator orientation, the position should be such that PORT and STBD on a vessel are orientated to the left and right of the operator respectively;

c) The Supplier should undertake a qualified HF analysis into the design of all compartment layouts, console configurations, maintenance envelopes, panel designs, HMIs and HCIs;

d) The Supplier should provide space and facilities for office administrative tasks within the control room;

e) The Supplier should ensure that eye-balling and verbal comms to other crew members is achieved in all key compartments through detailed Task Analysis and Link Analysis;

f) The Supplier should ensure that standing maintenance access to cabinets allows a minimum of 600 mm in front of that cabinet, including extra space for pullout racks;

g) The Supplier should ensure that crouching maintenance access to cabinets allows a minimum of 1200 mm in front of that cabinet, including extra space for pullout racks;

h) The Supplier should ensure that there is sufficient space for briefing at Incident Board locations commensurate with the briefing tasks of the crew;

i) The Supplier should ensure that one compartment design manager is in charge of each key compartment as a whole and takes responsibility for all HF design issues to ensure an integrated compartment CONOPS. This should include console layout, design and configuration, seating, lighting, ventilation and siting of Large Screen Displays (LSD) and ancillary equipment;

j) The Supplier should ensure that the ambient lighting is selected and designed to be compatible with the readability of panel information and vice-versa and that task lighting is provided at desk level where required through task analysis. Control rooms with windows should cater for day and night VDU readability, especially important on a ship's bridge where night vision is crucial for navigation and collision avoidance;

k) The system should be designed to operate within the temperature and humidity range appropriate for a manned compartment, usually about 20° C–22° C and 40% – 60% humidity. Standing and more active operators may require a cooler environment;

l) A sufficient number of communications devices should be available throughout the plant in order that users can report incidents quickly and effectively.

3.3.2 Design Outline Consoles

From the candidate outline console configurations, select the configurations that best meet the functional needs of the control room, e.g. single- or multi-screen, seated, standing, shared screen information, shared eye-balling, etc. The rules for console design are very similar, irrespective of the configuration, but some workstations are likely to need special features to suit the tasks of its host operator, for example, propulsion levers in a ship machinery control room, reactor control levers in a nuclear power station control room, crash stop fuel flow on an oil rig, etc. The following diagrams show a 1940s Lancaster Bomber cockpit (Figure 3.10) against a modern Boeing 787 jet airliner cockpit for dual operators designed to provide displays and controls optimally sited whilst maintaining the required outside fields of view (Figure 3.11). Note the recent innovation of flip down HUDs (Head-Up Displays) in the Boeing Dreamliner 787.

During the early development phase of the revolutionary 787 Dreamliner, the flight deck team was tasked with developing a design solution that would enhance safety while balancing innovation, cost, and operational commonality with previous Boeing flight decks. These key drivers were kept in the forefront as the team of engineers, pilots, and human factors experts took the initial flight deck concepts from paper layouts, to foam and cardboard ergonomic mock-ups, to functional prototype simulators, to final validation in full-scale hardware flight simulators.

Figure 3.10 Seated twin operator workspace for a 1940s Lancaster.

Figure 3.11 Seated twin operator workspace for a modern jet.

Throughout the process, the design decisions and prototypes were shared with customer pilots and regulators for their input to help ensure that the key design drivers were properly balanced. As a result of these efforts, the 787 flight deck incorporates a number of innovative features that offer opportunities for airlines to save money and enhance their competitiveness while maintaining operational commonality with the 777.

The following features of this Boeing jet cockpit are noteworthy because this 'control room' is probably one of the best designed, ever, (with Apple and Mercedes-Benz probably the best designers of user interfaces):

a) Primary flat panel electronic displays are centrally mounted with shared information between the two pilots, for example, engine health.
b) Primary controls are centrally mounted with shared controls between the two pilots, for example, thrust levers.
c) Critical display information is backed up by independent devices, for example, artificial horizon, airspeed, etc.
d) Critical controls are directly accessible, easily recognized and are shape-coded, for example, flight controls, gear selection, flaps, fire extinguishers, etc. Most importantly each pilot can see what the other is doing when 'in control' so that there is no doubt about the pilot's actions and the imminent behaviour of the aircraft, especially under emergency, reversionary or high workload conditions (see Chapter 5.15.2 on Automation). This has vitally important relevance to safety.

e) Controls and displays that are functionally related are generally grouped together except where specific tasks demand a grouping of devices that are not directly functionally related, for example, standby compass and fuel/fire devices. Task-oriented grouping of devices is recommended where this is more compelling than a functional-oriented grouping.

f) Both seats are designed specifically for this cockpit layout. For example, height adjustment aligns the pilots' eye levels to the same correct level for optimum field of view; fore and aft adjustment allows the pilots' feet to reach the pedals to apply the necessary forces required during asymmetric flight. Both of these adjustments are possible, even for pilots of widely different anthropometric dimensions.

g) In automated modes, for example 'Autopilot Engaged', each pilot can easily see what the autopilot is doing and how it is manipulating the controls. This is vital during emergency scenarios where rapid and unambiguous intervention is required, should the autopilot need to be overridden, for example, during a stall.

h) A HUD (Head up Display) collimated at infinity is provided to assist the approach during landing so that the pilots do not need to take their eyes off the runway.

3.4 Workspace and Panel Design

3.4.1 Design Workstations and Console Layout

Layout design should rely heavily upon the information generated from task analysis. This will affect how the workstation and panels are structured in terms of display/control relationships, grouping, sequences, etc. The three key criteria used when siting controls and displays are frequency of use, duration of use and importance of the devices:

1) **Single operator standing console** – showing standing eye position, limits for reach, desk height and angular positions for displays and controls (Figure 3.12).

2) **Single operator seated console** – showing seated eye position, limits for reach, desk height and angular positions for displays and controls (Figure 3.13).

3) **Console** Control and Display Locations – showing seated eye position and upper/lower limits for displays and controls (Figures 3.14–3.17).

4) **Multiple Control Room Locations** – very often, military operations demand that teams work together from various different control rooms on different platforms, typified by Re-fuelling At Sea (RAS), seen below conceptually between a Royal Fleet Auxiliary (RFA) tanker and a TYPE 45 destroyer (Figure 3.18).

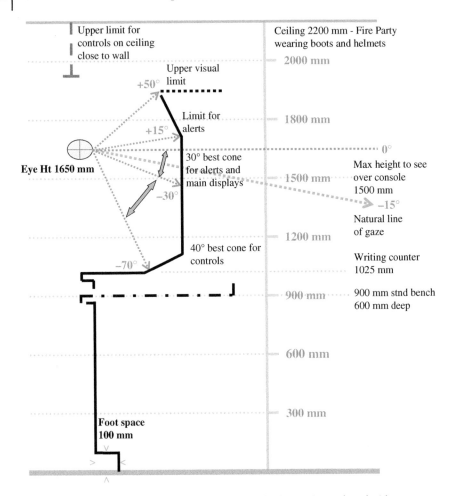

Figure 3.12 Example single operator standing console. *Source*: Reproduced with permission of Liveware HF Ltd.

3.4.2 Design Workstations and Consoles

Layout design should rely heavily upon the information generated from task analysis required to facilitate see-over (Figure 3.19):

Console Detail - Consoles should be designed to the following level of detail:

1) Sufficient freedom of movement for legs and a relaxed position of the trunk can be attained if the work surface is as thin as possible, maximum 30 mm;
2) If operators need to see over the top of the console then a low-profile design is necessary;
3) Most consoles require generous horizontal desk area for books, notes, writing, coffee cups, etc.;

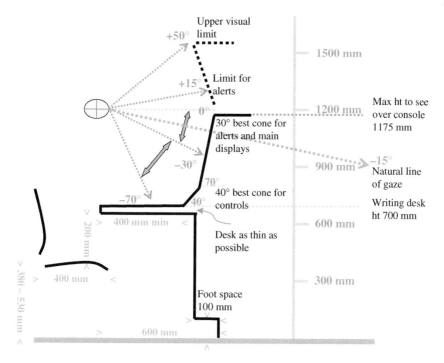

Figure 3.13 Example single operator seated console. *Source*: Reproduced with permission of Liveware HF Ltd.

4) Essential displays should be located within the operator's primary cone of vision as shown in Figure 3.13;

5) Frequently used controls that are connected with safety should be located in a centrally mounted position relative to the seated or standing workstation;

6) Controls should be located to the left or right of the console display in order to permit comfortable arm manipulation and leave free desk space for keyboards or writing directly in front of the user;

7) The system should ensure that control/display disassociation is avoided by generally locating controls next to their related displays;

8) Controls used in a strict sequence should be in either a column or row;

9) The console should provide dedicated hard-wired critical controls such as crash stops, fuel cut-offs, flight controls, etc.;

10) The console controls should be operable with all appropriate PPE (Personal Protective Equipment);

11) The console should be designed for a user population ranging between the 5th percentile female and 95th percentile male;

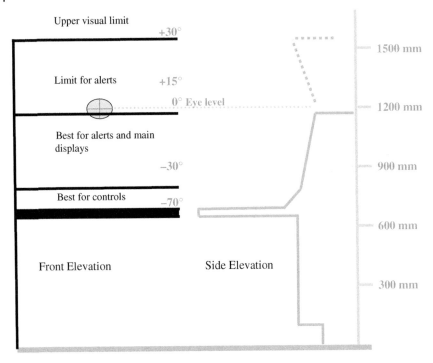

Upper visual limit
+30°

Limit for alerts   +15°

0° Eye level

Best for alerts and main
displays
−30°

Best for controls
−70°

Front Elevation   Side Elevation

1500 mm

1200 mm

900 mm

600 mm

300 mm

Figure 3.14 Approximate control and display locations. *Source*: Reproduced with permission of Liveware HF Ltd.

Figure 3.15 Example panel layout. *Source*: D. Bremner 2012. Reproduced with permission of ABB group.

12) Consoles should be integrated within the control room design and include horizontal desk areas for laying out documents and other hard copy items;

13) Consoles should provide sufficient adjacent storage for documents, paperwork, pens, stationery and beverage cups.

Figure 3.16 Example control room of a moving grate incinerator for municipal solid waste, Steag, Germany. *Source*: Published with the permission of VGB Power Tech GmbH Germany.

Figure 3.17 Traditional warship SCC control room circa 1980. *Source*: Reproduced with permission of Liveware HF Ltd.

Figure 3.18 Multiple control room operations across two different platforms. *Source*: MoD/ Crown copyright 2016.

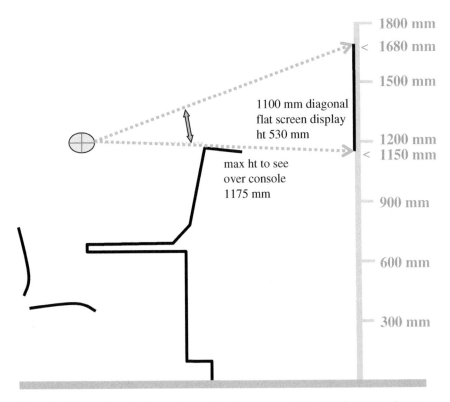

Figure 3.19 Example seated console requiring large screen display (LSD) viewing. *Source*: Reproduced with permission of Liveware HF Ltd.

3.5   Seating

The Supplier should provide a detailed description and diagrams of the seating, including dimensions and design features. This should be provided as an integrated part of the workstations and console to which it pertains.

Console Seating - The following HF principles should be considered for seating:

1) The seat should be designed in conjunction with the workstation (Figures 3.20 and 3.21).
2) The operator's back should be well supported, especially in the lumbar region.
3) The seat should be adjustable in height such that any operator within the 5th percentile female to 95th percentile male can attain a comfortable position. Designers should take note of the need for operators to establish the correct eye level with respect to the interfaces around them.
4) Seats should accommodate over-the-shoulder viewing and side-by-side seating by supervisors and for training purposes. Fixed armrests are not recommended.
5) Seating should support the need for Visual Display Unit (VDU) monitoring of more than one screen, for example, by providing a swivelling seat.
6) The seat covering should dissipate body heat and moisture. Leather breathes well, is durable, aesthetically pleasing and allows a degree of slippage to facilitate access and egress.
7) All seating should be integrated into the design of the console.

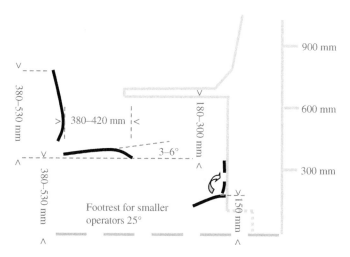

Figure 3.20 Approximate seat dimensions (elevation). *Source*: Reproduced with permission of Liveware HF Ltd.

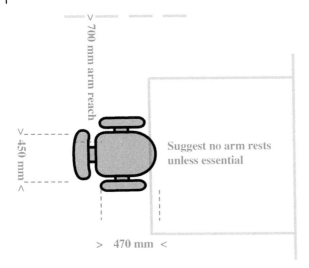

Figure 3.21 Approximate seat dimensions (plan). *Source*: Reproduced with permission of Liveware HF Ltd.

8) All seating should be designed using accepted ergonomic principles and be fully adjustable for swivel (lockable) fore/aft movement, height, backrest and cushion (the term 'squab' is used in the motor industry) angle, from the seated position.

9) The seat should be fixed to the deck or floor on moving platforms to ensure stability during motion.

3.6 Mock-ups and Example Workspaces

Simple mock-ups are inexpensive and highly effective in assessing console and workspace designs. This Liveware mock-up (Figure 3.22) was assembled and adapted from readily available office desks and used in early design iterations for the T45 SCC through MOD MES Abbey Wood:

The following photograph (Figure 3.23) features a simple and effective ergonomic mine hunter bridge, notably:

1) A light matt panel background to avoid reflections, provide an airy uplifting environment and enable clear functional positioning of major displays and controls;

2) Durable and well supported seat with plenty of adjustment, including a foot rest;

3) Good see-over capability for vessel manoeuvering.

Figure 3.22 Simple mock-up console for trialling a twin-screen display console. *Source*: Reproduced with permission of Liveware HF Ltd.

Figure 3.23 Mine hunter bridge panel layout. *Source*: Reproduced with permission of Liveware HF Ltd.

Figures 3.24 and 3.25 show early design work to ensure an ergonomic console design for the Queen Elizabeth aircraft carrier. It was notably important to ensure a good see-over capability of the multi-screen console displays onto the flight deck below:

Figure 3.24 Early aircraft carrier FLYCO configuration. *Source*: Reproduced with permission of Liveware HF Ltd.

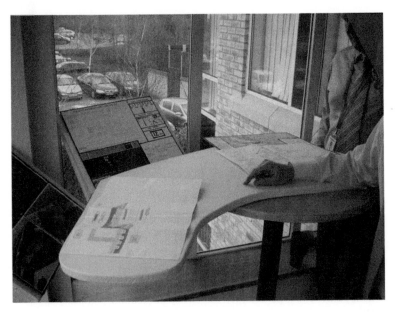

Figure 3.25 Early FLYCO console design allowing see-over. *Source*: Reproduced with permission of Liveware HF Ltd.

3.7   Maintenance

The Supplier must provide evidence of control room design features that take account of space for maintenance. Tasks required to implement maintenance must be identified so that the key spatial dimensions can be quantified.

The following examples represent some of the important dimensions:

1) Standing space required to access a cabinet at chest level should be 600 mm;
2) Crouching space required to access a cabinet or skirting at floor level should be 1200 mm;

3.8   Co-location

In a multi-control room building or platform, the Supplier must write a CONOPS for each control room outlining how it will operate in terms of intended equipment design, operator tasks, watchkeeping routines and personnel hierarchy. From this information, the relative locations of each control room can be determined from task-oriented data and thus ensure optimised co-location. Example factors that might determine co-location include:

1) Similar compartment functions;
2) Close verbal communication;
3) Information sharing.

4

Human-machine Interface Design

4.1 Outline Design Approach

This chapter covers the derivation of a sound Human-machine Interface (HMI) operating philosophy, that is, the detailed workstation, console, control and display design, arranged around any centrally mounted multi-function displays and the console seating (Figure 4.1).

4.2 HMI Operating Philosophy

The HMI philosophy should be based on the operators' manning, roles and tasks required in order to fulfil the control room functions. The purpose of an HMI Operating Philosophy is to facilitate the early development of design features, processes and requirements, some of which are mandatory, as to how the owner needs the plant and equipment to operate. This must occur within the boundaries of any legal obligations and constraints together with any mandatory health and safety standards in ensuring optimum operator safety, performance and comfort.

An Operating Philosophy provides a high level operational perspective in the use of the control room for controlling and monitoring systems, machinery, plant and equipment and provides operational details on all modes of control and surveillance. This document recommends that the HMI philosophy is derived on behalf of the plant owner and then provided to the owner's candidate Suppliers with the information needed to develop effective control room HMIs.

Human Factors in Control Room Design: A Practical Guide for Project Managers and Senior Engineers, First Edition. Tex Crampin.
© 2017 John Wiley & Sons Ltd. Published 2017 by John Wiley & Sons Ltd.

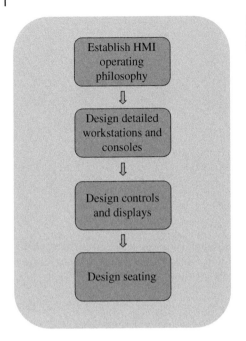

Figure 4.1 HMI design philosophy.
Source: Reproduced with permission of
Liveware HF Ltd.

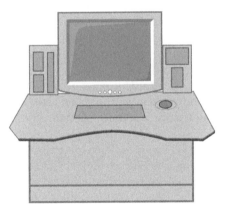

Figure 4.2 Single screen workstation.
Source: Reproduced with permission of
Liveware HF Ltd.

4.3 Detailed Workstation and Console Design

Detailed workstation and console HMI design principles follow on from the outline console designs covered in Chapter 3 above and should cover single and multi-screen assemblies.

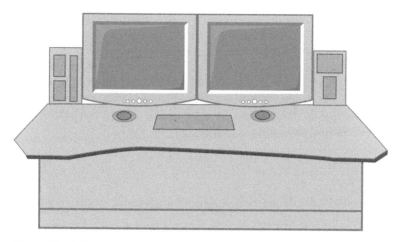

Figure 4.3 Multi-screen workstation. *Source*: Reproduced with permission of Liveware HF Ltd.

4.3.1 Single and Multi-Screen Workstations

Suppliers should provide a description of the single screen workstation hardware (Figure 4.2) and the multi-screen workstation hardware (Figure 4.3).

The workstation description and diagrams should include the following information:

1) Design and layout diagram with detailed dimensions, such as screen angles and height, work surface and storage provided, controls and spacing for operators sitting next to each other.
2) Integration with other equipments e.g. communication systems (internal), hard wired controls etc.
3) Display screen technology, including screen size and format ratio, screen resolution, brightness/contrast controls and range, etc.
4) Processor speed, graphics card, memory, etc.
5) Ruggedness, such as screen mounting and shock mountings (if required).
6) Weight.
7) Seating arrangement, notably allowing adjacent operators enough space to conduct their tasks without obstruction or allowing a single operator to access information from multiple screens. Sometimes it is useful to have two operators with three screens whereby the middle screen provides shared information.
8) Operational envelope around the workstation, allowing for seating, cleaning, traffic flow, maintenance access and egress.
9) Grab mounts, book storage, cup holders and pen storage, etc.

For a single operator at a twin screen console, the distance between the twin screen displays should be minimised. There should be a continuous flow between screens when using the cursor. The reversionary capability/procedure for converting a twin screen into a single screen should be described in the event of a screen failure. The Supplier should provide graphical mimic display illustrations for all of the systems where required, notably plant and machinery process graphics displayed for each main system and sub-system. These should follow the design principles outlined in Chapter 5 for the HCI.

4.4   Controls and Displays

The following broad rules should be used when laying out displays and controls:

1) Frequently used controls, or controls that are related to safety, should be located in a centrally mounted position relative to the seated or standing workstation;
2) Controls should generally be located below displays with free desk space for keyboards, main control devices, documents or writing space directly in front of the user;
3) Ensure control/display association by locating controls next to their related displays where possible;
4) Controls used in a strict sequence should be installed in either a column or row;
5) Controls should be operable with all appropriate PPE (Personal Protective Equipment), e.g., anti-flash gloves;
6) Design for a user population ranging between the 5th percentile female and 95th percentile male so that the smallest can reach objects and the largest have clearance beneath objects;
7) Ensure that all seating is integrated into the design of the console, be it seated or sit/standing;
8) Ensure that all seating is designed using accepted ergonomic principles and is fully adjustable for swivel (lockable and stable on a moving platform) fore/aft movement, height, backrest and squab (cushion) angle, from the seated position;
9) Provide consoles that are integrated within the compartment design and include horizontal desk areas for laying out hard copy items such as technical references, charts and other paper-based items;
10) Ensure that consoles provide sufficient storage for documents, paperwork, pens, stationery and beverage cups;
11) Ensure that display screen real estate and the benefits of multi-screen consoles are decided on the merits of the information required by the user for one, two or more VDUs through task analysis;
12) Ensure that all consoles have a see-over capability where users need eye contact with their colleagues or to wall mounted LSDs.

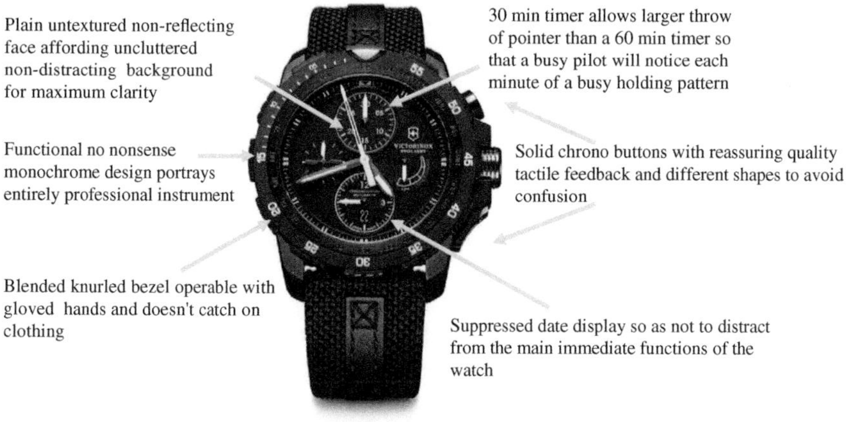

Plain untextured non-reflecting face affording uncluttered non-distracting background for maximum clarity

Functional no nonsense monochrome design portrays entirely professional instrument

Blended knurled bezel operable with gloved hands and doesn't catch on clothing

30 min timer allows larger throw of pointer than a 60 min timer so that a busy pilot will notice each minute of a busy holding pattern

Solid chrono buttons with reassuring quality tactile feedback and different shapes to avoid confusion

Suppressed date display so as not to distract from the main immediate functions of the watch

Figure 4.4 Well designed functional watch developed for the Swiss Army. *Source*: Reproduced with permission of Liveware HF Ltd.

The following small device (Figure 4.4) has been chosen as a good example of display/control design in a wrist watch with features explained around the dial. This watch has a no-nonsense military appearance with a beautifully clear dial, avoiding the common distractions typical of many watches (with over complex functionality, poor contrast, poor readability and awkward usability). The watch is quite heavy with a deep case which can cause discomfort when worn over prolonged periods. The outer bezel, when used for timing, counts down from 60 instead of up; this could cause confusion for a busy operator since the stop watch function counts up as it should do. If it was thinner, quartz and made of titanium (as many watches are now) it would be one of the best designs around.

4.4.1 Large Screen Displays

The Supplier should provide a description of the Large Screen Display (LSD) hardware. The LSD description and diagrams should include the following information:

1) How the display is controlled – LSDs should not be ganged to the primary overview pages of a workstation such that the information changes without warning. What is displayed on a LSD can be controlled from a workstation but the pages displayed on the workstation should not automatically be replicated on the LSD without careful consideration;
2) Display screen technology including screen size and format ratio, screen resolution, brightness/contrast controls and range;
3) Processor speed, graphics card and memory;

4) Ruggedness, such as screen mounting and shock mountings (if required);
5) Weight.

4.4.2   Interactive Large Screen Displays

Where interactive (smart) LSDs are proposed by the Supplier, their hardware characteristics should be described and how the HMI varies from the non-interactive LSDs.

4.4.3   Palmtops

Where palmtops are proposed by the Supplier, their hardware characteristics should be described together with a description of the tailored HCI.

4.4.4   Pagers

If pagers are to be provided, the Supplier should provide a detailed description of the pager hardware characteristics, including how the pager operates, how information is presented/controlled and how the operators respond to its alerts.

4.4.5   LOPs (Local Operating Panels)

Many control rooms coordinate information from geographically separated locations containing local equipment such as pumps, gas turbines, diesel engines, etc. Some of these local equipments have their own panels for local control which should be designed to follow a similar operating philosophy as those remote controls in the main control room. The Supplier should provide a detailed description of the Local Operating Panel (LOP) hardware characteristics.

4.4.6   Hardwired Controls

The Supplier should provide a detailed description of hard wired controls including where they are located on the various workstations and consoles and how they are used. For example, hardwired controls should include:

1) PDLs (Power Demand Levers);
2) Crash stops;
3) Alert controls (e.g. accept, mute);
4) Trips;
5) Critical/frequently used function buttons;
6) Screen brightness/contrast controls;
7) Communications.

Operator interaction using the intended devices should be tested during trials using realistic scenarios.

4.4.7   Fire (and Flood) Detection Panels

The Supplier should provide a detailed description of the fire (and flood) detection panels. The description should include:

1) Location in compartments of the fire panel, (and flood panel if appropriate), detector heads, etc.
2) These panels should interface both with the system and display system diagrams on a graphical mimic display, for example, all the detectors (heat, smoke and flame) in the different compartments around the plant.

4.4.8   Fire Suppression Panel

The Supplier should provide a detailed description of the fire suppression system including its location in all rooms.

4.4.9   CCTV

The Supplier should provide a detailed description of the CCTV system, including displays and controls. The description should include:

1) Details of the number and location of CCTV cameras around the plant with supporting justification;
2) Details of how the CCTV information is presented, preferably on separate dedicated VDUs. The Supplier should work with the Design Authority to agree how the information is presented and to what level of detail objects need to be seen.

4.4.10   Printers

The Supplier should provide a detailed description of how printers are integrated into the control rooms in order to be spatially compatible with the system operation, notably:

1) Minimising printer noise in order to maintain the 55 dBA office noise levels;
2) Ensuring printer access without compromising traffic flow.

4.4.11   Reversionary Modes of Operation

All critical systems should have some form of reversionary operation in case of failure or emergency. The Supplier should undertake a risk assessment in order to ascertain where reversionary modes are required and how they should be implemented.

4.5 Alerts (Alarms and Warnings)

4.5.1 Alerting Philosophies

The main Alerts design areas include:

1) A classification of Alerts:
   a) **Alarm** – needs operator response within seconds, eg, GT tripped, fire in main plant;
   b) **Warning** – needs operator response within minutes, eg, lub oil temperature high, reactor level rising;
   c) **Events** – normal system operations that require monitoring but not necessarily an operator response, e.g. clutch disengaged, pumps running;
   d) **Messages** – notes between operators, e.g. a report following a test.
2) Why Alarms and Warnings are needed;
3) Alarm and Warning standards and guidelines;
4) Alarm and Warning reduction techniques;
5) Alarm and Warning console design issues including:
   a) Workstations and HCI devices;
   b) Presentation and handling;
   c) Audible Alarms and Warnings;
   d) Messages, events and notifications.

The Supplier should provide a detailed description of how operators respond to Alarms and Warnings, sometimes collectively known as 'Alerts'. The presentation and handling of the control and display information should be provided in detail. This should include the following Alert topics:

1) Classification;
2) Transferable sub-system Grouping;
3) Group status indicators (Alert borders);
4) Filtering;
5) Displays;
6) Summary Alert info;
7) Alert list page;
8) Summary Alert list;
9) Alert information;
10) Alert tabs;
11) Colour coding.
12) Paging;
13) Navigation buttons;
14) Handling Alerts;
15) Annunciation;
16) Logs.

The Supplier should provide details of the Alerts to be provided in terms of operation, look and feel and deviate only with prior written agreement with the Design Authority. Issues to be covered where relevant include:

1) Alert philosophy;
2) Structure and prioritisation;
3) Groupings;
4) Reduction techniques;
5) Types of Alerts:
   a) Alert;
   b) Warning;
   c) Events;
   d) Messages;
   e) System connection lost.
6) Types of display:
   a) Summary list in banner;
   b) Alerts page;
   c) Stateboards.
7) Presentation:
   a) Summary information;
   b) List behaviour (e.g. Alert enters at top/bottom of list, speed of entry, movement when an Alert is hooked, etc.);
   c) Alert content information (e.g. date and time stamp, message, high/low limits etc.);
   d) Flash rates;
   e) Colour coding;
   f) Symbols;
   g) Categorisation tabs (systems, role based, SIC, alarms vs warnings).
8) Auditory Characteristics:
   a) types;
   b) tone;
   c) pitch;
   d) pulse;
   e) loudness;
   f) timbre;
   g) muting;
   h) direction;
   i) voice output.
9) Hyperlinks to system pages;
10) Scrolling mechanism;
11) Handling and Controls:
    a) hooking;
    b) acceptance;
    c) mute;
    d) hardwired.

4.5.2 Design of Alerts

Reaction - Operators react 25 ms quicker to auditory alerts in a quiet environment than visual alerts.

Inhibiting - Alerts that are activated when equipment is started or shut down, in its normal state, should be inhibited. The operator should only have his attention drawn to alerts which are out of the ordinary, for example, fire detection.

Hierarchy - Alerts should occupy an initial hierarchy so that they can be filtered and categorised at a later date in the design.

Aviation - Flight Deck experience shows that there is a need to prioritise an alerting system into four categories:

1) **Class 1 Alert** – departure from safe flight profile, e.g. stall, ground proximity: colour code red;
2) **Class 2 Alert** – abnormal configuration, e.g. landing gear failure, low fuel: colour code amber;
3) **Class 3 Alert** – aircraft system status, e.g. VOR out of range: colour code blue, white or green;
4) **Class Alert 4** – communications, e.g. interphone failure: colour code blue, white or green.

General Categories - Alert Response Categorisation (Table 4.1):

Number: The maximum number of high significance alerts should **not be more than about 20 – 30** to maximise the operator's capabilities.

Benchmark Targets - Targets have been established for alert management in industrial control rooms. In the context of experience gained from major accidents in the offshore, nuclear and aerospace industries, these targets provide a useful insight into the targets that may be appropriate for warships. The industrial benchmark targets are summarised below:

1) Less than one significant alert every 10 minutes;
2) Less than 10 significant alerts in the 10 minutes following a significant occurrence;
3) Less than 10 standing alerts;
4) Less than 30 'shelved' alerts.

Specification: The following Table 4.2 demonstrates how a series of practical steps can be applied to the specification and design of systems so that operators receive a manageable number of prioritised alerts:

Management: Alerts should be managed within the following guidelines. Full annunciation details will be evolved in accordance with current HF guidelines:

1) Operators must be able to trace the cause of the alert, even if the alert condition was transitory and has cleared. The latching of alerts should be considered;

Table 4.1 Alert response categorisation.

Level	Title	Response	Implementation	Annunciation
1	Alert A1	Emergency	Immediate and significant threat to safety of system, major equipment or personnel. Initial operator actions must proceed without further investigation.	A1 Visual. Alert tone – high frequency dual tone: 5 Hz red.
2	Alert A2	Immediate	Implications of condition are immediate and severe but operator intervention may depend on further evaluation.	A2 Visual. Alert tone - high frequency monotone - 5 Hz red.
3	Warning W1	Priority	Operator must respond to condition at first opportunity if alert or similar event is to be avoided.	W1 Visual. Warning tone - high frequency monotone: 2 Hz amber.
4	Warning W2	Urgent	Operator attention needed to avoid condition but no inherent threat to integrity of power plant.	W2 Visual. Warning tone - med frequency monotone: 2 Hz amber.
5	Warning W3	Investigate	Operator must be made aware of condition. Urgency of response subject to operator assessment	W3 Visual only - low frequency monotone : 2 Hz amber
6	Warning W4	Low Priority	Operator attention appropriate as soon as opportunity Arises.	W4 Visual only - low frequency' monotone: 2 Hz amber.
7	Warning W5	Non-Operational	No direct threat to ongoing operation of the plant or equipment. Issue can be addressed in due course by non-watchkeepers.	Information available on request. Operators made aware that Level 5 warnings in force - low frequency monotone: 2 Hz amber.
8	Event	Information Only	Notifies the operators of an occurrence that does not imply any transition towards an alert or warning condition.	All events to be logged. Associated information available to operators.

Source: Reproduced with permission of Liveware HF Ltd.

2) The annunciation of alerts is to be unambiguous and easily distinguishable from all other types of alert;

3) Since alerts, by definition, imply emergency actions, the system should provide appropriate checks and procedures to help ensure that the correct action is taken.

Table 4.2 Alert specification and design process.

Step	Objective	Technique
1	Avoid conditions where alerts are generated.	a) Establish operating parameters within which plant can follow routine transients without crossing alert boundaries; b) Enable plant to predict conditions where alert will be generated and take avoiding action.
2	Allow alerts under specific conditions to be detected, interpreted and resolved by plant without passing alert to operators.	a) Identify transients caused by acceptable sets of circumstances and over-ride alert; b) Enable automatic response to alert condition and hide from operators unless further parameter boundaries (time or value based) are crossed; c) Enable auto-inhibits; d) Enable manual inhibits.
3	Maximise time within which operators must respond to an alert.	a) Specify systems with sufficient capacity, redundancy or local protection to avoid excessive transients; b) Enable operating modes that will delay the need for operator intervention; c) Incorporate automatic responses to mitigate problem until considered action can be taken.
4	Prioritise alerts so that operators are not distracted by minor events.	a) Distinguish between several layers of alerts and warnings.
5	Combine multiple alerts into single alert for initial actions, allowing multiple alerts to be resolved in slow time.	a) Where multiple alerts provide redundant information (e.g. multiple trips for loss of power supplies could be combined into single alert for the associated loss of capability).
6	Apply filtering.	a) Apply logical filtering; b) Apply AI (neural network based) techniques.
7	Identify least intrusive annunciation.	a) Use audio alerts only where essential: vary tone and volume; b) Vary style of visual alerts; c) Consider position and type of displays used.
8	Allocate alerts to interested parties only.	a) Allocate alerts to watchkeepers/maintainers/command/trainees etc. as appropriate.

Source: Reproduced with permission of Liveware HF Ltd.

Technical Issues: Where appropriate, the issues discussed below are to be considered for all types of alert but they are best addressed in the context of warnings (since alerts tend to be rigidly defined responses to specific conditions):

1) Provision is to be made for warning levels to be adjusted by those with appropriate authorisation;

2) An appropriate and adjustable range of hysteresis must be provided so that parameters do not go into warning repeatedly for minor fluctuations in parameter level. For example, a tank level sitting close to the warning level should not generate repeated warnings because of the normal motion in a seaway for a marine system;

3) Auto-inhibit functions need to be provided. These are a central part of the strategy to minimise unnecessary alerts. Sufficient attention must be paid during the design phase to achieve a comprehensive and reliable set of inhibits that will prevent warnings being generated by plant outside of the defined operating conditions. Comprehensive auto-inhibits can involve complex logic and some of the applicable techniques will form part of the wider options for alert 'filtering';

4) Manual inhibits need to be provided, for example, lube-oil pressure on an engine which would be low until the revs increase. Operators must be able to identify when machinery is in an abnormal condition and follow simple procedures to eliminate potential alerts within specified limits. Close attention must be paid to establishing when manual inhibits will be cancelled by a change in the status of the plant. If manual inhibits are automatically cancelled by the system, the inhibit configuration must be stored in memory and re-instated with ease if the original conditions return. In this respect, operators should have access to a database of inhibits so that previously used inhibits, and their constraints, are readily available;

5) One option is for manual inhibits to be supported by a graphical representation of their logic so that operators can construct and review the logic tree being applied. The aim is to make the manual inhibit facility more powerful and ensure that the logic is immediately visible for management scrutiny;

6) Events are to be readily available to the operators but are not to distract them from responding to alerts;

7) Events are to be subject to appropriate recording and analysis. In addition to the basic collation of parameter data, histories and trends, any dedicated analysis for specific events is to be identified as early as possible in the design process.

Design Recommendations: Recommendations can be summarised as follows:

1) Present attentional directors close to the operator's line of sight within a maximum 15° cone for high priority alerts and 30° cone for all other warnings;
2) Use a master signal within a 15° cone of the line of sight if attentional directors cannot be placed within a 30° cone;
3) Eliminate the possibility of confusing alerts with any other type of display;
4) Ensure alerts are presented until the operator has responded or until the alert state is no longer active;
5) Ensure alerts have at least twice the luminance of other displays in the working environment;
6) Use larger characters for alert text (up to about 60 minutes of arc), especially under adverse viewing conditions;
7) Accepted alerts should be clearly distinguishable from unaccepted alerts;
8) Consider polarity changes of contrast for different alert text and symbols;
9) A sound and flashing light should accompany the onset of an alert;
10) On accepting an alert, any accompanying sound should cease and the flashing resort to a steady light;
11) Add auditory voice signals to high priority alerts;
12) Auditory alerts and warnings should be presented at a sound level above the normal ambient noise of the equipment or control room;
13) A pulsed tone is more likely to be received by an operator than a continuous tone;
14) Female voices are more likely to be obeyed by male operators;
15) Perception of red information is about 75% slower than the perception of other information in peripheral vision since colour perception is poorer peripherally;
16) Signals presented in the bottom of the visual field (about 60 minutes below the line of sight) are detected slightly faster than those at the top;
17) Depending on size and contrast, close alternately flashing stimuli in peripheral vision can appear as one moving stimulus, caused by an effect sometimes referred to as stroboscopic apparent motion. Two alerts may be perceived as only one, hence all flashing indicators should be synchronized;
18) Auditory signals should not be expected to carry the detailed information that is contained in words until attention has been gained. Once attention has been gained by an initial nonverbal signal or attention, it is permissible to add spoken words later;
19) Presentation of alerts should lead the operator through the perceptual (initial detection), decision making and psycho-motor (action required) processes concerned with handling that alert;

20) In the absence of any sophisticated alert handling system, alerts should be listed chronologically. Most recent alerts should appear at the bottom of the list;

21) For listed alerts, the entire background colour, red or amber, should flash. Note that the text should be either white or black in order to achieve the best contrast ratio for readability;

22) Scrolling up or down needs to be at a speed that enables either rapid access to an alert some distance up or down the list, or fast access while still being able to read each alert as it scrolls by;

23) Inactive alerts should be accessible for training purposes but not as part of normal operating procedures;

24) Alerts that have been accepted should change from flashing to steady. Once dealt with, they should disappear from the list. Should an alert not be accepted and it subsequently resolves itself, for example, by a low pressure resuming its normal state, then this should still flash until accepted, whereupon the alert disappears;

25) A mute should be provided for some alerts. However, this should be confirmed by trials appropriate to the design issue;

26) A separate accept button should be provided for alerts;

27) The first operation of the 'alert accept' button should mute all current alerts. Note that each alert has to be accepted individually in order to stop it flashing. The first operation of the 'warning accept' button should mute all current warnings. Warnings can be accepted either individually or as a whole to economise on time;

28) The individual alert that is to be accepted should be separated slightly from the other alerts listed in order to make it clear which alert is being accepted.

Coding: For alert display coding:

1) Coding of alerts and warnings should follow accepted human stereotypes, e.g. red for alarms, amber or yellow for warnings and green for normal running. Military users expect certain stereotypes (accepted formats) for alerts, the most common being the Red – Amber/Yellow - Green convention (RAG). Here Red, Amber and Green are used to indicate Alarm, Warning and Normal conditions;

2) Use flash coding sparingly for directing attention;

3) Where flash coding is used, ensure that:

   a) The background is free of flashing stimuli;

   b) All flashing alerts are synchronised at between 2 and 10 Hz (4 Hz is preferred) with a duty cycle of 70%;

   c) Text flashes at no more than 2 Hz, with a duty cycle of 70%.

Philosophy: Generally, all alert systems should operate within an agreed philosophy that will vary depending upon the application. An example philosophy might be as follows:

1) Alerts should be presented separately but adjacently on a vertical time history with the most recent alert at the bottom.
2) Time should be resolved to within 1 second.
3) Text should include a clear description of the plant item and the nature of the problem, for example, 'Port Generator Temp High', 'Fire in No. 2 Engine', 'Reactor Critical', etc.
4) The individual alert accepted should be separated slightly from the other alerts listed in order to make it clear which alert is being accepted.
5) When dealt with, alerts should disappear from the list.
6) Operators should use the recent history of alerts to assess plant trends. The presentation of previous alerts should facilitate the efficient extraction of information by the operators and the search for any particular type of alert should be as effortless as possible. Some alerting systems often collate past alerts into long, monotonous tables which operators have to scan repeatedly to extract relatively simple information. In many cases, it will be helpful if the display or page layout allows operators to compare selected information in parallel.
7) In diagnosing from historical alerts, operators may want to look at two windows of alert information. This should be considered in screen real estate calculations.
8) **Flash rates** – 2 Hz for warnings and 5 Hz for alarms.
9) **Scroll** – a speed that enables rapid access to an alert some distance up or down the list together with smooth continuous scrolling at a readable pace.
10) **Accepting** – Alerts should remain on the alert list until the alert condition has cleared and the alert has been acknowledged.
11) The system needs to be provided with a means to archive all system data including, but not limited to, the alert and event logs, dynamic data and alert set points.
12) Alarm and Warning pages need to be large and scrollable.

Primary Alerting: is shown below (Figure 4.5) by example from the TYPE 23 Frigate whereby alerts on the actual permanent mimics illuminate for ease of detection and association.

Secondary Alerting: shown below whereby alerts on an embedded VDU illuminate but are only visible if that page is selected. The alerts are shown by red and yellow borders around the items in question in order to directly associate the alert with the object (Figure 4.6).

Glass Cockpit Approach: Modern warships are rapidly acquiring display systems that have been regularly used by the air lines for many years. MoD

Figure 4.5 Alerts in the TYPE 23 Frigate Illuminated in yellow over the permanent mimics.

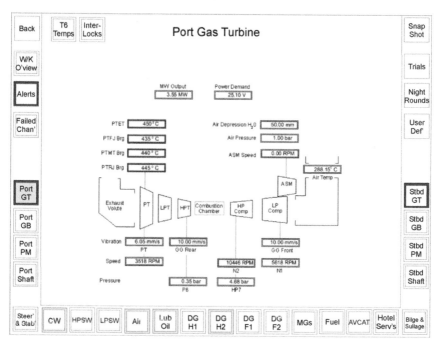

Figure 4.6 Generic prototyping of alerts for the Royal Navy. *Source*: Reproduced with permission of Liveware HF Ltd.

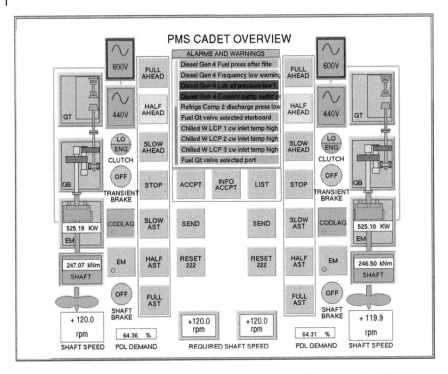

Figure 4.7 Generic prototyping (circa 1998) of a touch screen HCI based on the TYPE 23 frigate. *Source*: Reproduced with permission of Liveware HF Ltd.

research into Platform Management Systems (PMS) in the 1990s put forward a touchscreen 'glass cockpit.' Trials revealed a high level of satisfaction from RN users provided that the touch target areas on the screen were large enough, typically 30 mm x 30 mm for gloved hand operation (Figure 4.7).

5

Human-computer Interface Design

5.1   Outline Design Approach

This chapter covers the derivation of a sound Human-computer Interface (HCI) operating philosophy, that is, the detailed control and display design arranged around any centrally mounted multi-function displays (Figure 5.1).

5.2   General HCI Operating Philosophy

5.2.1   Introduction

The HCI operating philosophy starts with a CONOPS (Concept of Operations) which provides an outline description of how operators will use a system in responding to various events. A CONOPS should describe in broad terms who does what, when and where and be tested through prototyping using operational scenarios. The operating philosophy should be based on the operators' manning, roles and tasks required in order to fulfil the control room functions. The purpose is to facilitate the early development of design features, processes and requirements, some of which are mandatory, as to how the owner needs the plant and equipment to operate. This must occur within the boundaries of any legal obligations and constraints together with any mandatory health and safety standards in ensuring optimum operator safety, performance and comfort.

An Operating Philosophy provides a high level operational perspective in the use of the control room for controlling and monitoring systems, machinery, plant and equipment and provides operational details on modes of control and reversionary modes of operation. This document takes the HMI philosophy derived in Chapter 4 above on behalf of the plant owner and provides candidate Suppliers with the information needed to develop effective control room HCIs.

Human Factors in Control Room Design: A Practical Guide for Project Managers and Senior Engineers, First Edition. Tex Crampin.

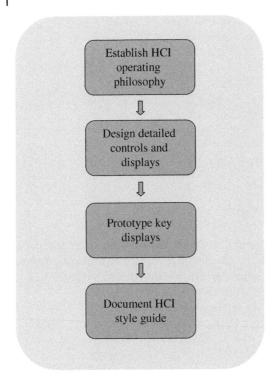

Figure 5.1 HCI design philosophy. *Source*: Reproduced with permission of Liveware HF Ltd.

5.2.2 General HCI Design Principles

The following list summarises the key factors required in developing detailed electronically paged HCIs which, for the moment, is the most likely solution for the near future:

1) Under office ambient lighting conditions, display screens should have a light and neutral background. Under low level ambient lighting conditions, display screens should have a darker and neutral background.
2) Page titles, critical information and the primary navigation bar should be permanently available.
3) Hyperlink buttons should be provided within the display pages to provide the operator with quick access to critical sub-pages elsewhere in the page hierarchy.
4) Alert borders should be provided around the main navigation components to lead the operator through pages in the case of an alert or warning.
5) Screen components that enable system control should be clearly identifiable and shown by a raised appearance (or other coding dimension). Conversely, status information should appear flat so that it is clear what can be selected and vice versa.

6) Overview pages should be provided where appropriate, to pull together the key information available at lower levels in the page hierarchy, for example, where role-based workstations are required for watchkeepers.

7) Displays should be readable when viewed away from the perpendicular up to 45°, especially for displays read whilst on the move.

8) Important text should be presented at a size of 20–25 arcmin to the user's eyes at the normal working location, seated or standing. This equates to about 1% of the viewing distance so as a rule of thumb key text labels should be about 6 mm high at 600 mm. As the viewing distance increases so must the text size, proportionately. Less importance text can be smaller and major headings or titles larger.

9) Important symbols (graphics such as pumps, breakers, valves, etc.) should be presented at a size of 45–50 arcmin to the user's eyes at the normal working location, seated or standing; This equates to about 1.5% of the viewing distance so as a rule of thumb key symbols should be about 9 mm high at 600 mm. As the viewing distance increases so must the symbol size, proportionately. Major symbols such as main plant items must be even larger and provide a balanced appearance on the screen.

10) Display readability must be compatible with the ambient lighting and task demands of the user. For example, on a ship's bridge, displays must be readable at night and in full sunlight. Such is the dynamic range of ambient outside lighting that displays for these applications must have at least two modes; a daytime mode and a night mode. Further, this requires the development of different colour palettes so that sufficient brightness and contrast is achieved between screen objects. Generally, a daytime screen will have a light background and a night screen a dark background.

11) Primary display information should be presented to the user within a 30° cone of view, that is, 15° either side of the normal resting eye line-of sight which is 10° below the horizontal.

12) Console displays (monitored for more than a few seconds) should be aligned to this normal resting line-of-sight, usually 10° below the horizontal. It is very tiring to monitor displays above the horizontal for long periods.

13) Where operators control multiple functions, a minimum twin-screen display should be considered in order to ensure situational awareness and the ability to zoom into specific system detail in parallel.

14) Provide mimics that accurately reflect the layout of plant systems.

15) All mimics should be uncluttered and portray only sufficient objects for the operator to access and control the information required to undertake the task. Simply copying circuit diagrams or system process charts is unlikely to provide an optimum HCI.

16) Unless there are compelling arguments not to, mimic displays should generally represent the spatial position of equipment in relation to other rooms, zones or compartments.

17) The system should provide a dedicated mimic that displays the activation of fire alert detector heads in each area.

18) The system should provide an intuitive dialogue by which users navigate around multi-page displays and obeys HF rules of consistency and ease-of-use.

19) The system should not allow HCI mimic displays to be modified in any way unless subject to formal design change procedures by the Authority. This is especially important for colour coding, size and shape of standard objects which must be consistent across the plant HCIs.

20) The system should provide immediate access to critical information, displays and controls through hardwired non-mimic displays, i.e., gauges for operating parameters of primary systems. The instrumentation to be hardwired should be agreed with the Design Authority.

21) The system should provide an HMI/HCI that is flexible and sympathetic in handling any likely errors made by the user and should allow for the correction of errors by the user, e.g. undo/reset functions.

22) Navigation paths through mimic displays should be traceable to the top level display thus allowing the user to see the context of any current view in terms of its relationship and relevance to the whole system. For this reason menus should be wide and shallow in order to avoid getting lost in deep menu structures.

23) The system should enable rapid task closure, that is, rapid completion of one task before going on to the next task and, clear feedback that the first task has been completed.

24) The system should provide a mimic display that makes it clear where the user has control, e.g. where a cursor can facilitate a control action.

25) The system should provide instant feedback to users that a control input has been actioned and accepted;

26) The system should provide instant display of important requested information, typically within 200 ms.

27) The system should display all alerts instantly through a dedicated Alarms and Warnings mimic that is always available.

28) Menu hierarchies should be wide and shallow, not narrow and deep.

29) The system should allow the user to return to the previous page in one single action.

30) Animation should be avoided unless it is essential for task feedback otherwise it becomes a distraction.

31) There should be a consistent colour scheme for status information throughout the system.

32) The system should provide delineation of the different system functions if presented on the same mimic display.

33) The system should provide sufficient numbers of screens (screen real estate) for each operator by task analysis of key information needed under worst case conditions.

34) The system should monitor and display the activation of key machinery items, e.g. closing down vents or activation of fans for smoke clearance, etc.

35) Where information is to be displayed on an alternative VDU display (e.g. laptop) other than the main displays, it should be legible on the smaller screen and follow the same operating philosophy.

36) The system should provide feedback to the user through clear visual and auditory processes where critical automated tasks are involved, e.g. operation of main circuit breakers, starting of gas turbines, reactor shut-down, etc.

37) The system should be designed with consideration for the orientation and configuration of the mimic displays to represent the layout of rooms and compartments around the facility.

38) Visual indications should be provided when plant or equipment are running, e.g. vent running indications and crash stop controls.

39) The system should enable ease of navigation around the displays through shortcuts and back page buttons (both hard buttons and on-screen soft buttons).

40) Ensure that any mirror imaged consoles for left and right, Port and Starboard, front and back, forward and aft locations undergo rigorous HF scrutiny to ensure consistent and safe usability. Mirror imaged panels can be confusing to operators under certain circumstances.

5.3 Detailed Design of Controls and Displays

5.3.1 Introduction

First, it must be recognised that optimised control and display design starts with a thorough understanding of the workspace, operator roles, HCI screen real estate, procedures and system functions. Software designers must understand how their endeavours will be used to establish usable HCIs.

Suppliers should demonstrate that their software developers understand how the software within the HCI element of a graphical mimic display is to be used and applied in the specific control room concerned. This should be achieved by the writing of brief Operability Statements that cover the control room design from compartment level down to the actual HMIs/HCIs as follows:

1) **Functions** – all of the features of the system in terms of what it does;

2) **Compartments** – a functional and physical description of the compartments including illustrations of how they may be operated and manned at each readiness state;

3) **Operator Roles** – the human resources required to carry out the compartment functions in terms of their roles, rank, tasks, qualifications and training needs;

4) **HCIs** – including an early assessment of screen real estate;

5) **Communication** – different levels of communications facilities that the operators may employ;

6) **Standard and Emergency Operating Procedures** – the specific routine and emergency tasks required to monitor and control the plant;

7) **Systems** – a functional and physical description of the systems, e.g., HPSW system, reactor primary loop, main drilling rig, etc., and illustrations of how each system might be used in all recognised states or conditions.

5.3.2   Functional Software Overview

Most systems are typically built up of key functional software components in order to achieve monitoring and control. The diagram below (Figure 5.2) shows an example relationship between functional software components for a Warship Platform Management System (This will be generically similar to any major plant but obviously providing different functions).

The remaining advice in this chapter focuses on the actual design of the HCI itself.

5.3.3   Specific HCI Design and Layout Rules

The Supplier should implement the following display page HF Operability Statements:

1) On a Warship, all VDUs used in spaces subject to a 'Darken Ship' procedure should be NVG (Night Vision Goggles) compatible.

2) The Supplier should ensure that all displays are integrated into the control room layout (i.e. positioning, height, etc.) as a whole to ensure user operability.

3) All mimics should be uncluttered and portray only sufficient objects for the operator to access and control the information required to undertake the task, to be agreed with the Authority.

4) All mimic displays should accurately represent the position of equipment in relation to the area and level of the plant, rig or ship. These should always be visible as a less prominent greyed out or dotted background to the key display mimic information.

5) The HCI should provide each sub-system on one mimic display with the ability to view individual or lower level information in more detail, e.g. high pressure sea water, electrical distribution, etc.

6) Where data is known to be estimated it should be clearly identified on the mimic display as being an estimated reading/display, e.g. estimated tank contents.

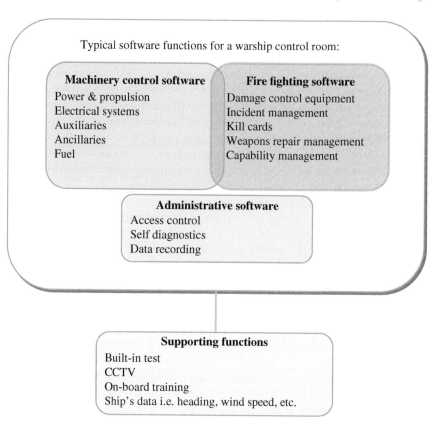

Figure 5.2 Example functional software overview. *Source*: Reproduced with permission of Liveware HF Ltd.

7) The system HCIs should be consistent throughout the plant.

8) The system should provide an intuitive dialogue by which users navigate around multi-page VDUs and obeys system-derived HF rules of consistency and ease-of-use.

9) Any differences in intuitive dialogue (accepted normal human stereotypes) due to plant operations or user characteristics should be carefully considered. Sometimes it is necessary to violate accepted design norms because of plant idiosyncrasies. Thus, if there is anything about specific plant procedures that demands an HCI that deviates from a standard ergonomic design, then this should be considered and implemented very carefully. An example is the DI (Direction Indicator) in an aircraft cockpit. The DI usually deploys a moving rotating scale against

a fixed pointer which is generally disliked in the HF standards documentation. However, because of the very special nature of piloting an aircraft spatially, using points of the compass, it works very well and is universally accepted as a good design solution. The same moving scale fixed pointer design is sometimes used, confusingly, on domestic central heating controls. The user rotates the dial anti-clockwise to increase the temperature, violating an accepted human stereotype. Conversely, rotating bath taps clockwise decreases the flow of water. Whilst this also violates an accepted human stereotype it is so universal, and has been around for so many years, that everyone is comfortable with this control solution.

10) The system should allow equipment parameters to be changed through the mimic display only by authorised personnel, e.g. temperature sensing levels on a diesel generator.

11) The system should provide immediate access to critical information, displays and controls through hardwired non-mimic displays, i.e. gauges for operating parameters of primary systems.

12) The gauges to be hardwired should be agreed.

13) The system should provide an HMI/HCI that is flexible and sympathetic in handling any likely errors made by the user.

14) The system should allow for the correction of errors by the user, e.g. undo/reset function.

15) The system should provide instant display of requested information.

16) The system should continuously monitor all sub-system parameters.

17) The system should provide a mimic display that uses zooming only for specific functions that would benefit from this HCI feature, e.g. for an electronic state-board showing fires and floods in various areas. Further, rather than actively zooming, in most cases, it is better to zoom to the required detail in one step.

18) The system should provide mimic displays that have key user information that is permanently on all display screens, e.g., management aims across the top of the screen.

19) The system should provide sufficient numbers of screens in key compartments.

20) The system should provide feedback to the user through clear visual and, where appropriate, auditory feedback processes where critical automated tasks are involved, e.g. operation of main circuit breakers.

21) The system should provide an HMI/HCI to visually indicate when systems, plant and equipment are running, e.g. vent running indications and crash stop controls.

Example graphical mimic display pages are shown below (Figure 5.3):

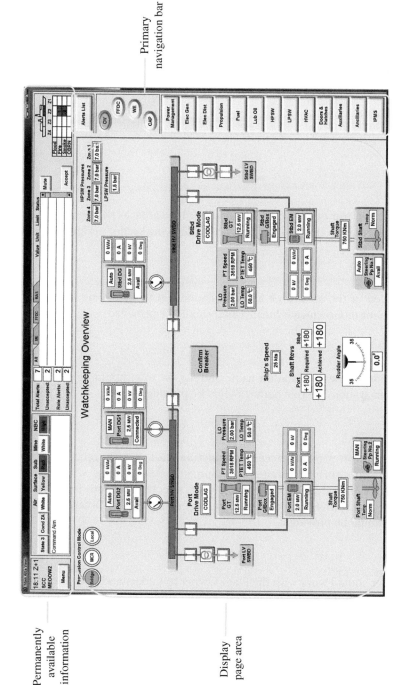

Permanently available information

Primary navigation bar

Display page area

Figure 5.3 Example overview of HCI components on a warship platform management system (PMS). *Source:* Reproduced with permission of Liveware HF Ltd.

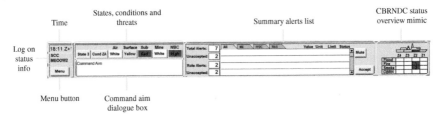

Figure 5.4 Example of permanently available information at top of screen. *Source:* Reproduced with permission of Liveware HF Ltd.

5.3.4 Permanently Available Information

The area at the top of the screen should be dedicated to the permanently available information that is described in the following sections (Figure 5.4).

5.3.5 Time

An indication of the local time and its relationship to Zulu time (Universal Time Constant or Greenwich Mean Time) should be provided.

5.3.6 Log-on Status

An indication of the workstation name and log-on role should be provided. A list of workstation and role names should be provided.

5.3.7 Menu Button

The menu button provides access to functions not directly related to the control and monitoring of systems/equipments. Menu items will include Logging On/Off, watch or shift changeover, selecting Station–in–Control, changing to and from the night palette, print screenshot, etc. Details of these functions should be provided.

5.3.8 Conditions and Threats

The Supplier should work with the Plant Owner to understand how their staff conduct fire fighting and damage control and understand damage control conditions, the operating conditions and the different states. The Supplier should display the current plant state, conditions and levels.

Illustrative examples of how the operator changes the conditions should be provided. An indication of which roles have the authorisation to perform these functions should be provided.

5.3.9 Command Aim Dialogue Box

A Command Aim Dialogue box should be provided to display the current Command, Mission or Plant Aim. An indication of what roles have the authorisation to perform these functions should be provided.

Illustrative examples of how an operator changes the Command Aim should be provided.

It is recommended that when the Command Aim change is made it should flash locally until accepted by the operator at each workstation. Once actioned, all Command Dialogue boxes should be updated throughout the plant. Whether an indication that it has changed is provided at the workstations needs to be addressed, as it is assumed that a main broadcast on the tannoy will be made in parallel.

5.3.10 Summary Alerts List

A summary alerts list should be provided. A description of how this is presented and operated should be provided.

5.3.11 Damage Control Status Overview Mimic

A simplified profile mimic, showing the damage control areas should be provided. Blocking the relevant zone with a solid colour should indicate alerts. There should be a distinction between accepted and unaccepted alerts (e.g. accepted alert steady, unaccepted alert flashing indication status). There should be a table beneath the graphic showing the total number and types of active alerts within each zone.

5.3.12 Additional Permanently Available Information/Controls

The provision of the following indications/controls should be considered:

1) **Training Mode Indication/Control** – if a training mode is to be provided an indication as to whether normal operating mode or training mode is currently selected needs to be provided. A description of how this feature works should be provided including selection/de-selection processes, additional mode coding, etc;
2) **On-Line/Off-Line Indicator** – an indication whether the workstation is connected correctly to the server and network should be provided.

5.3.13 Display Page Area

This area of the screen holds the Display Pages that are selected from the primary navigation bar and hyperlinks/secondary navigation bars within the display pages. It is in this area that the operator should monitor and control (if authorised) the system related equipments.

In designing the graphical mimic overview pages, the Supplier should ensure that:

1) There is a fixed space for the system, plant or machinery being displayed and that within this fixed space other windows cannot be opened;
2) Child screens within the main page area are not used;
3) Pop-up windows are used to a minimum;
4) The graphical pages are designed with thought to high tempo and stress situations so that any functionality that allows the operator to configure the screen to suit their preference is kept to a minimum;
5) Zooming is kept to a minimum.

5.3.14   Primary Navigation Bar

The Primary Navigation Bar is permanently available and divided into three main parts:

1) Alerts list button;
2) Navigation group buttons;
3) System navigation buttons.

The three parts are described in more detail in the following sections, including example pictures.

5.3.15   Alert Button

This provides access to a full page of alerts within one button press. Selecting the button will call up the Alerts List page in the Display Page Area.

5.3.16   Alerts

The following Alerts HF Operability Statements are mandatory for the Suppliers to implement:

1) Group alerts should be made available at other key workstations for plant faults. Group alerts for specific systems should be grouped separately, e.g., shut down of main machinery;
2) The system should instantly display new alert information on all display screens;
3) All alerts should be displayed in the order they occur.

5.3.17 Navigation Group Buttons

The Navigation group buttons, see adjacent example, should provide access to the following pages:

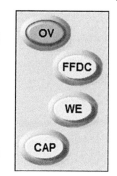

1) Watchkeeping overview page;
2) Fire Fighting and Damage Control pages;
3) Weapons Engineering related pages;
4) Capability Management pages.

Selecting a button should call up the relevant page in the Display Page Area.

5.3.18 System Navigation Buttons

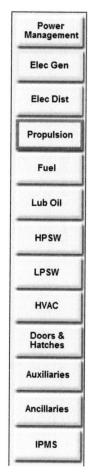

The system navigation buttons (see adjacent example) should provide access to the key system pages relevant to the operator role (within one button press). This part of the Primary Navigation bar is generally ordered to reflect the importance of the systems (most important at the top) and groups functionally related systems together.

Suppliers should change the system navigation buttons depending on which functional group the operator is most interested in, e.g. main plant system engineering, fire fighting and damage control, etc. If this philosophy is adopted the Navigation group buttons and some Secondary Navigation bars (within the Display pages) would need to be modified accordingly. Thus the mandatory requirement is that this prioritisation process is undertaken with SME input.

For example, the system navigation buttons indicated to the right are primarily of interest to the Supervisor's role. Other operators are likely to be less interested in some of these systems and more interested in FF and DC systems. The illustration to the right indicates that a component within the electrical generation system has a warning associated with it (yellow border) and that a component within the propulsion system has an alarm associated with it (red border). This will help the operator to navigate through the display pages to help identify the cause of the alerts. This philosophy should be used throughout the system for all navigation buttons where appropriate. This philosophy will only work if spurious alerts are eliminated; prototyping is required to resolve the alerting philosophy.

Secondary navigation bar

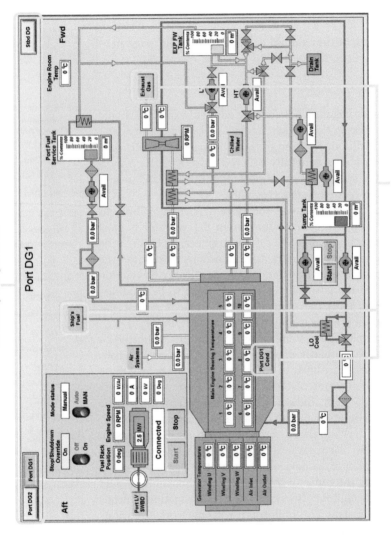

Hyperlinks

Figure 5.5 Example secondary navigation bars and hyperlinks. *Source:* Reproduced with permission of Liveware HF Ltd.

5.3.19   Secondary Navigation Bars and Hyperlinks

In addition to the Primary Navigation bar, display pages should be called up in the Display page area by using:

1) Secondary Navigation bars – the preferred method (Figure 5.5);
2) Hyperlinks – should be kept to a minimum.

5.3.20   Secondary Navigation Bars

The Secondary Navigation bar should be located at the top of the display page and allow access to the top level display pages within that group. Selecting a button will call up the relevant page and make the button appear depressed.

If the operator selects another page from the Primary Navigation bar and then later returns to the group of pages, the last viewed display page will be automatically displayed. For example, if the operator selects the 'Auxiliaries' button from the Primary Navigation bar and then selects the 'Chilled Water' button from its Secondary Navigation bar, the 'Chilled Water' page will be displayed. If the Operator then selects the 'HPSW' button from the Primary Navigation bar, the 'HPSW' page will be displayed. If then the operator reselects the 'Auxiliaries' button from the Primary Navigation bar the 'Chilled Water' page will be automatically displayed.

5.3.21   Hyperlinks

Hyperlinks should be located within the Display pages, providing quick access to related systems in another group within the page hierarchy.

Generally if a hyperlink is provided within one page, once operated, there should be a corresponding hyperlink to take the operator back to the page previously viewed.

5.3.22   Types of Display Page

The following types of overview, system and surveillance display pages are provided as examples of the different types of display design formats required:

1) Overview pages (Figure 5.6);
2) System pages: split design (Figures 5.7, 5.8);
3) System pages: control panel/mimic design (Figure 5.9);
4) Secondary surveillance pages;
5) Ring main mimic page (Figure 5.10);
6) Night colour palette pages (Figure 5.45);
7) Stateboard pages (not shown for security reasons).

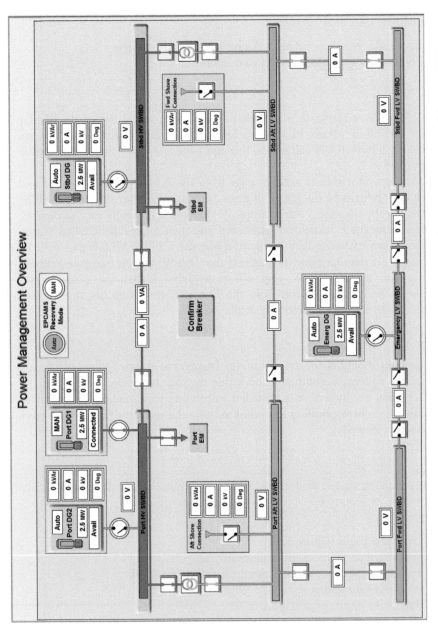

Figure 5.6 Example overview page: power management overview. *Source:* Reproduced with permission of Liveware HF Ltd.

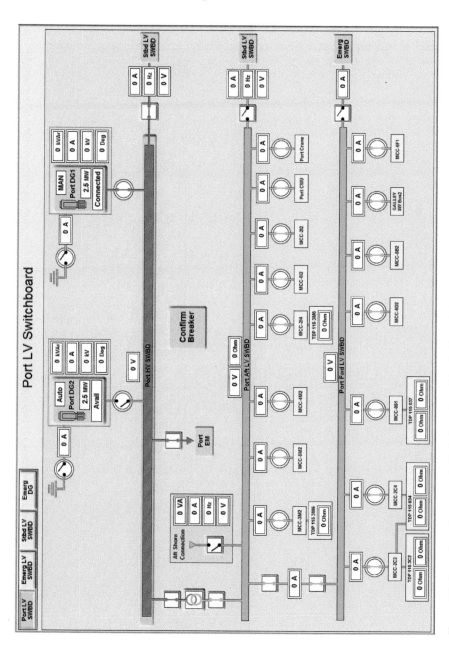

Figure 5.7 Example system page: Split design – Port LV switchboard. *Source*: Reproduced with permission of Liveware HF Ltd.

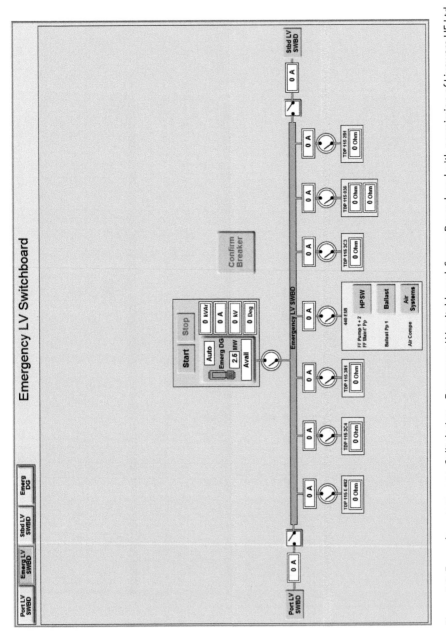

Figure 5.8 Example system page: Split design – Emergency LV switchboard. *Source:* Reproduced with permission of Liveware HF Ltd.

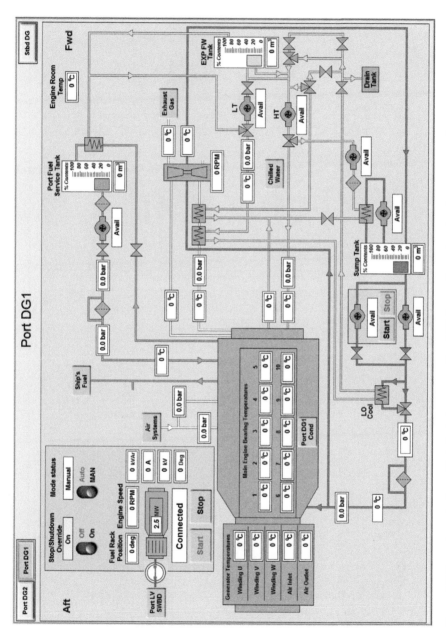

Figure 5.9 Example control panel/mimic design page – Port DG1. *Source*: Reproduced with permission of Liveware HF Ltd.

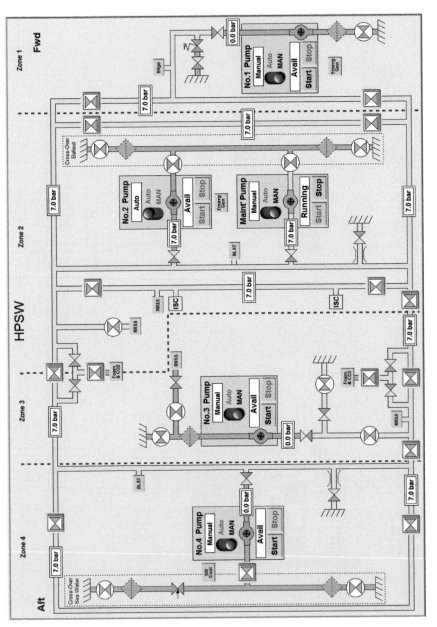

Figure 5.10 Example ring main mimic page – HPSW. *Source:* Reproduced with permission of Liveware HF Ltd.

5.3.27   Night Colour Palette Pages

Night and day colour palettes should be provided for system screens used in compartments with windows (Figure 5.45). An example display page indicating the implementation of a night colour palette should be provided.

5.3.28   Stateboard Pages

An example display page indicating the presentation format of a Stateboard page should be provided.

5.3.29   Single Screen Navigation and Control Philosophy

The single screen HCI operating philosophy is applicable to all single screen workstations and laptops (Figure 5.11).

The single screen operating philosophy is consistent with that of the left hand screen of the twin screen philosophy. The following design features should be applied:

1) The Primary Navigation bar is positioned on the right hand side of the screen;
2) The single pointer on the screen is controllable by a trackball;
3) Control of equipment is permissible on any of the display pages if the system allows control of that equipment and the operator has Station-in-Control.

Two single screen workstations can be sited side by side and used as a twin screen workstation, but a single operator would have to use two different trackballs to control the two cursors.

5.3.30   Twin Screen Navigation and Control Philosophy

The twin screen HCI operating philosophy should comprise two displays mounted side-by-side using one trackball/pointer to operate both screens (Figure 5.12). The twin screen displays are 'Independent' from each other in that different display pages can be displayed on each screen. All twin screen workstations provided should have the same Navigation and Control philosophy.

The left hand screen should display the same permanently displayed information as that on a single screen HCI. The right hand screen should display additional permanently available information at the top of the screen (e.g. HPSW info and system information). A possible alternative could be to have a 'scratch pad' that allows the operator to write notes relevant to the watch or equipments (similar to the function of a 'post-it' note).

Both of the screens should have a Primary Navigation bar; the left hand screen should have its bar on the right and the right hand screen should have its bar on the left. The following design features/issues should be noted:

1) Control of equipment is permissible on any of the display pages on any of the screens if the system allows control of that equipment and the operator has Station-in-Control;

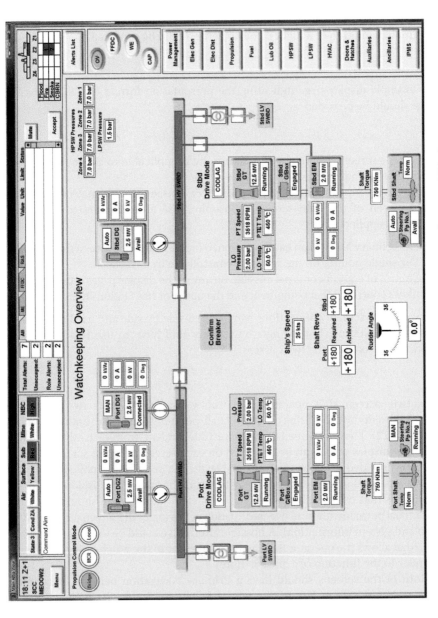

Figure 5.11 Example single screen HCI. *Source:* Reproduced with permission of Liveware HF Ltd.

Permanently available information

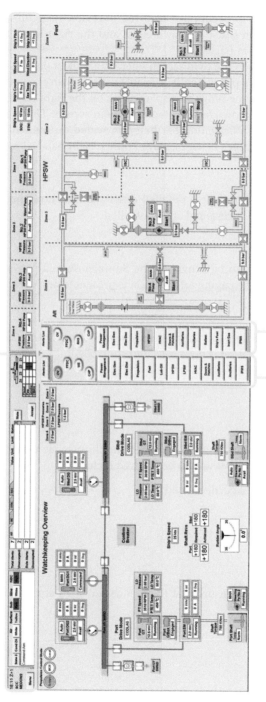

Left hand screen primary   Right hand screen primary
navigation bar   navigation bar

Figure 5.12 Example twin screen HCI. *Source*: Reproduced with permission of Liveware HF Ltd.

2) Operators should procedurally keep the overview page on the left hand screen to maintain situational awareness and the right hand screen to call up different system pages.

   (An alternative to the twin screen operating philosophy described above would be to have a permanently available overview display page on the left hand screen allowing only the display pages on the right hand screen to be changed. This is known as an 'integrated' twin screen. This would result in only one Primary Navigation bar being provided on the left hand screen that controls the pages being displayed on the right hand screen.)

5.3.31   Large Screen Display Navigation and Control Philosophy

The Supplier should provide a detailed description of how the large screen display information is presented and controlled.

5.3.32   Paging Operating Philosophy

If a paging system is to be provided, the Supplier should provide a detailed description of how the paging information is presented and controlled.

5.3.33   Palmtop HCI Operating Philosophy

If a palmtop is provided, the Supplier should provide a detailed description of how the HCI information is presented and controlled.

5.4   Menus

5.4.1   General

If menus are to be provided, the supplier should provide details in terms of operation, look and feel. Issues to be covered include:

1) How they are invoked/dismissed;
2) Shape;
3) Size;
4) Colour;
5) Highlighting;
6) Item selection.

   The supplier should ensure that the menus provided are easy to use with a trackball input device.

5.5   Windows

5.5.1   Page Windows

The supplier should provide a description, with examples, on how windows are opened and closed in concept.

5.5.2 Pop-up Windows

Where pop-up windows are used the supplier should provide details in terms of operation, look and feel. Issues to be covered where relevant include:

1) Opening and closing;
2) Re-sizing;
3) Moving;
4) Determining the opening location and size;
5) Scrolling;
6) Minimising the size of the window (iconising);
7) Shape, size and colour;
8) Title.

5.5.3 Pop-up Window Example

The diagram below (Figure 5.13) illustrates a pop-up window used to provide control of a DG:

1) The window is opened by right clicking on the associated component button (left click performs the hyperlink function). The window is closed by selecting the 'X' button in the top right hand corner;
2) The window appears in a precise location that maintains association with the component being controlled and does not obscure any other information;
3) The window title indicates the associated component and the type of function being performed;
4) In this instance it is not possible to resize, minimise or move the window and there is no need to provide a scrolling mechanism.

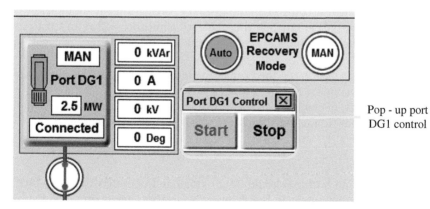

Figure 5.13 Example pop-up window. *Source*: Reproduced with permission of Liveware HF Ltd.

5.6   Controls

5.6.1   General Presentation

The supplier should provide details of control presentation in terms of operation, look and feel. Issues to be covered where relevant include:

1) Size, shape and profile;
2) Appearance – pointer over;
3) Appearance – selected;
4) Colour;
5) Labelling;
6) Trackball left click and right click;
7) Audible feedback.

All controls should activate their control function on left or right 'click- up'.

The following sub-sections provide examples of control descriptions. Generally all the controls appear as raised square/rectangular components (buttons), with the following exceptions:

1) Navigation group buttons;
2) Right clicking on some status information can open an associated pop-up window or menu;
3) Mode selection controls are raised but rounded in shape.

5.6.2   Navigation Controls

The following types of Navigation buttons should be provided:

1) Navigation group buttons;
2) System navigation buttons;
3) Hyperlinks.

Only one navigation button on the Primary Navigation bar should appear selected (depressed) at any one time; selecting another will return it to its 'up' state. The same philosophy is also true for the Secondary Navigation bars.

The navigation buttons on the Primary Navigation bar should be larger than those on the Secondary Navigation bar.

All text, icons, dialogue boxes etc. provided on navigation buttons, should move by one pixel to the right and down when the button is depressed.

5.6.3   Navigation Group Buttons

Navigation group buttons should be provided and operate with a left click. An example arrangement in the Primary Navigation bar is provided below

(Figure 5.14), the unique elliptical shape and format making the buttons stand out.

The font is 11 pt Arial Bold.

The Navigation group buttons should have the following presentation states (Figure 5.15).

5.6.4 System Navigation Buttons – Primary Navigation Bar

An example array of system navigation buttons on the Primary Navigation bar is provided below. The buttons are rectangular in shape and operated with a left click. They have the following arrangement (Figure 5.16).

The font is 11 pt Arial Bold.

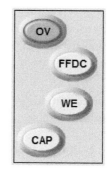

Figure 5.14 Example navigation group buttons: Arrangement. *Source:* Reproduced with permission of Liveware HF Ltd.

Figure 5.15 Example navigation group buttons: Presentation states. *Source:* Reproduced with permission of Liveware HF Ltd.

Presentation state	Illustration
Button up	FFDC
Pointer over while button up	FFDC
Button down	FFDC
Pointer over while button down	FFDC
Warning - Alert border activated	FFDC
Alarm - Alert border activated	FFDC

The system navigation buttons on the Primary Navigation bar should have the following presentation states (Figure 5.17):

5.6.5 System Navigation Buttons – Secondary Navigation Bars

System navigation buttons on the Secondary Navigation bars are rectangular in shape and operated with a left click. An example of their arrangement is provided below (Figure 5.18).

The font is 11 pt Arial Bold.

The system navigation buttons on the Secondary Navigation bars should have the following presentation states (Figure 5.19).

5.6.6 Hyperlinks

The Hyperlink buttons provided on the Display pages are, by example, rectangular in shape. Different sizes are used, depending on their importance and information provided on the button and the space available within the page. There are three main sizes. The largest size consists of a group of sizes, where the hyperlink buttons integrate status information. The medium and small sizes just contain the hyperlink label.

The buttons are operated with a left click and will only appear depressed momentarily before returning to their 'Up' state.

The hyperlink buttons will have the same presentation states as those provided for the system navigation buttons.

Some examples of the large and medium hyperlink buttons are provided below (highlighted with yellow border) (Figure 5.20).

Some examples of the small hyperlink buttons on the HPSW page are provided below (highlighted with yellow border) (Figure 5.21).

5.7 Machinery Controls

The supplier should provide a description of all types of machinery controls to be provided. Consistency of look, feel and operation should be maintained wherever possible. Examples are provided below:

5.7.1 DG Start/Stop Buttons

The DG Start and Stop buttons should be provided on DG related overview pages as pop-up windows and permanently on the DG pages as illustrated below (Figures 5.22, 5.23).

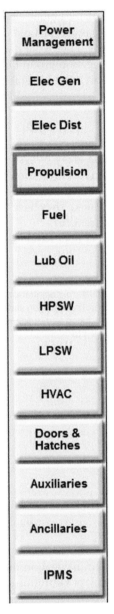

Figure 5.16 Example system navigation buttons: Primary navigation bar arrangement. *Source*: Reproduced with permission of Liveware HF Ltd.

Figure 5.17 Example system navigation buttons: Presentation states. *Source*: Reproduced with permission of Liveware HF Ltd.

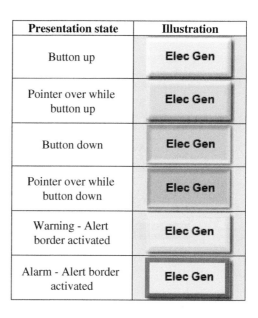

Presentation state	Illustration
Button up	Elec Gen
Pointer over while button up	Elec Gen
Button down	Elec Gen
Pointer over while button down	Elec Gen
Warning - Alert border activated	Elec Gen
Alarm - Alert border activated	Elec Gen

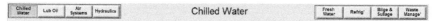

Figure 5.18 Example system navigation buttons: Primary navigation bar arrangement. *Source*: Reproduced with permission of Liveware HF Ltd.

Presentation state	Illustration
Button up	Hydraulics
Pointer over while button up	Hydraulics
Button down	Hydraulics
Pointer over while button down	Hydraulics
Warning - Alert border activated	Hydraulics
Alarm - Alert border activated	Hydraulics

Figure 5.19 Example system navigation buttons: Presentation states. *Source*: Reproduced with permission of Liveware HF Ltd.

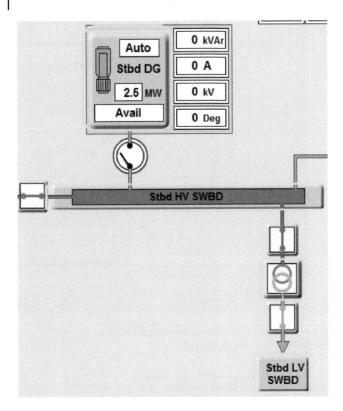

Figure 5.20 Example hyperlink buttons (large and medium sizes). *Source*: Reproduced with permission of Liveware HF Ltd.

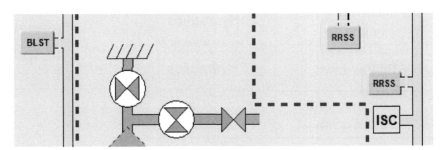

Figure 5.21 Example hyperlink buttons (small size). *Source*: Reproduced with permission of Liveware HF Ltd.

The Start and Stop buttons should be operated with a left click and will only appear depressed momentarily before returning to their 'Up' state. They have the following presentation states (Start button example provided) (Figure 5.24).

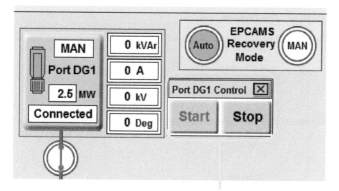

Pop-up port DG1 Start/Stop buttons

Figure 5.22 Example pop-up window start/stop controls: Power management overview page. *Source*: Reproduced with permission of Liveware HF Ltd.

Figure 5.23 Example permanently available start and stop controls on Port DG1 page. *Source*: Reproduced with permission of Liveware HF Ltd.

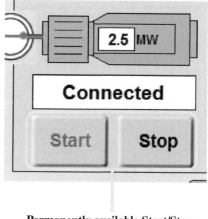

Permanently available Start/Stop buttons

(Consideration should be given to the need for a confirmation stage when starting/stopping the DG on the DG pages)

5.7.2 Mode Select Controls

Mode select controls allow the operator to select the required mode for an equipment and integrate the status information. With a two-way switch the operator can left click anywhere on the control to change to the alternate mode. The figures below illustrate the controls for two different modes associated with a DG (Figure 5.25).

Presentation state	Illustration
Button up	**Start**
Pointer over while button Up	**Start**
Button down	**Start**
Control available	**Start**
Control not available	Start

Figure 5.24 Example start button: Presentation states. *Source*: Reproduced with permission of Liveware HF Ltd.

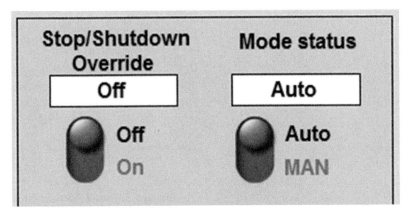

Figure 5.25 Example mode controls: Most common states. *Source*: Reproduced with permission of Liveware HF Ltd.

The most common mode is presented with the circular control located at the top (Figure 5.26).

The label associated with the control position is black when the control is next to it and grey when the control is in the alternate position. When the mode is set to Local, both the 'Auto' and 'MAN' labels and mode control will be greyed out (Figure 5.27).

(Consideration should be given to the need for a confirmation stage when switching between equipment modes)

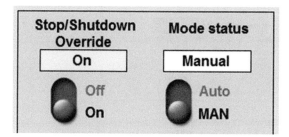

Figure 5.26 Example mode controls: Less common states. *Source*: Reproduced with permission of Liveware HF Ltd.

5.7.3 Breaker Controls

The breaker controls should be square buttons that are operated by a left click and integrate the status information. On left clicking, the button should momentarily appear depressed and then return to its up position.

Breaker controls are provided as discrepancy switches that allow the operator to select the state they want the breaker to go into (the discrepancy state) and then confirm the action. The breaker is operated by selecting it and then selecting the Confirm Breaker button within five seconds. If the

Figure 5.27 Mode control and labels greyed out. *Source*: Reproduced with permission of Liveware HF Ltd.

operator does not select the Confirm Breaker button within five seconds the breaker button will return to its true state and the Confirm Breaker button will become disabled. Selecting a breaker button that is in its discrepancy state will immediately return the breaker button to its true state (Figure 5.28).

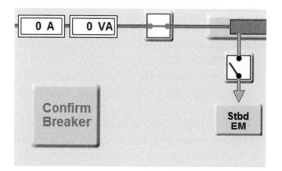

Figure 5.28 Example breaker controls: In true state (one in horizontal orientation and one in vertical orientation). *Source*: Reproduced with permission of Liveware HF Ltd.

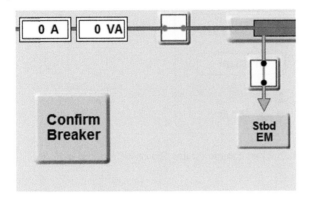

Figure 5.29 Example breaker controls: One breaker in discrepancy. *Source*: Reproduced with permission of Liveware HF Ltd.

Only one breaker should be operated at a time. Selecting a second breaker button will return the other breaker button in discrepancy to its true state (Figure 5.29).

Figure 5.29 illustrates breaker buttons and the associated Confirm Breaker button.

The breaker buttons should have the following presentation states (Figure 5.30):

The button should remain in the up state if control is not available, i.e. left clicking will not initiate the button down state. For example, this may occur if the breaker has no electrical supply. If control is not available when the breaker has malfunctioned the Alert border will be activated.

The text on the associated Breaker Confirm button should appear grey when the button is disabled and black when enabled (i.e. a breaker is in discrepancy).

5.7.4 Valve Controls

The valve controls are square buttons operated by a left click and integrate the status information. The button on left click should momentarily appear depressed and then return to its up position.

The figure below illustrates a valve control on the HPSW page (Figure 5.31):

The valve buttons have the following presentation states (vertical when open orientation shown) (Figure 5.32).

If operation of valves needs to be confirmed, they should work in exactly the same way as the Breaker Discrepancy controls, but with the alternative icons.

The position of the valve should be animated in relation to the valve's real position in order to show the discrepancy.

Figure 5.30 Example breaker button and icon: Presentation states (vertical orientation). *Source*: Reproduced with permission of Liveware HF Ltd.

Presentation state	Illustration
Button up	
Cursor over while button up	
Button down	
Breaker OOA	
Breaker live	
Breaker dead	
Breaker live but in discrepancy	 Flashes between the above 2 states (complete cycle every second)
Breaker dead but in discrepancy	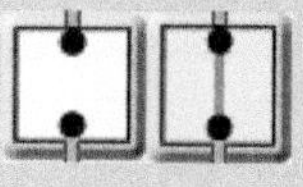 Flashes between the above 2 states (complete cycle every second)
Warning - Alert border activated	
Alarm - Alert border activated	
System connection Lost - Alert border activated	

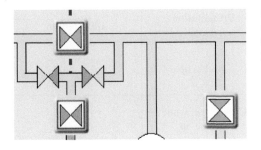

Figure 5.31 Example valve controls (two valves open and one valve closed). *Source*: Reproduced with permission of Liveware HF Ltd.

Presentation state	Illustration
Button up	
Cursor over while button up	
Button down	
Valve OOA	 OOA
Valve open	
Valve closed	
Warning - Alert border activated	
Alarm - Alert border activated	
System connection Lost - Alert border activated	

Figure 5.32 Example valve button and icon: Presentation states. *Source*: Reproduced with permission of Liveware HF Ltd.

Valves that have a graduated open/closed position should show that gradual movement in order to provide feedback to the operator on the precise valve behaviour.

5.7.5 Keyboard Controls

The keyboard should be of a QWERTY format unless a special HCI requirement dictates a different format.

In addition to hard wired controls, some controls may be provided on the keyboard, e.g. function keys. The supplier should provide a list of all the keys that perform specific functions.

5.8 Dialogue Boxes

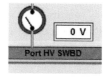

Dialogue boxes provide status information to the operator. Equipment states and variables that have an associated alert should be provided with an alert border (Figure 5.33).

The Supplier should provide descriptions of all of the various types of dialogue boxes to be provided.

Figure 5.33 Example dialogue boxes with alert borders. *Source*: Reproduced with permission of Liveware HF Ltd.

Some dialogue boxes are integrated into hyperlink buttons to save space (Figure 5.34).

5.9 Use of Colour

5.9.1 General Use of Colour

The following general HF rules apply for colour coding:

1) **Policy** – Document a colour policy across the whole system and adhere to it rigidly;
2) **Stereotypes** – Use colours that have standard human stereotypes, e.g. Red for fire or alert, amber for caution, green for go or normal, blue for water, etc.;

Figure 5.34 Example dialogue boxes integrated into hyperlink. *Source*: Reproduced with permission of Liveware HF Ltd.

3) **Number of Colours** – Keep this to an absolute minimum, about five, unless there is very good reason to need more. Code graphics in monochrome first, making full use of black, white and shades of grey before introducing colour later to enhance only the key features of the graphic;

4) **Redundant Coding** – Ensure that key features are coded both in colour and another coding dimension, e.g. shape (Figure 5.35);

5) **Background Colour** – Use a light neutral background (light grey) for displays used operationally in a daylight or brightly illuminated environment (office). Use a darker neutral background for twilight and night use (see Section 5.14 on day/night colour pallets). This technique is used successfully in many car satellite navigation displays;

Figure 5.35 Vehicle hazard warning. *Source*: Reproduced with permission of Liveware HF Ltd.

6) **Colour Palette** – Where possible, colour should be used sparingly in order to avoid visual confusion. Due to the expense of printing in colour, where colour is required but can only be printed in monochrome, the Microsoft Word RGB Colour Co-ordinates are provided to enable close to true reproduction. Note that colours on the screen will look different when printed off so must be tested in their final form. The following recommended 15 shades/colours are shown thus, all reasonably well distinguishable from each other. However, it is recommended that as few colours are used as possible and for many systems it is unlikely that all 15 colours would need to appear simultaneously on one page. For most non-specialised systems four shades and four to five colours should be sufficient. Note the need to change the text from black to white for certain background colours to be legible as shown in Figure 5.36.

5.9.2 Specific Uses of Colour

The following specific HF Rules apply for colour coding of RN warship and submarine machinery control room mimics on multi-colour electronic displays:

Pastel Colours – Use pastel colours in order to avoid clashing with the necessarily more saturated Green (Running), Alarm (Red) and Warning (Amber/Yellow) colours. A colour regime is shown above (Figure 5.34) for a warship machinery control room display with black text on a grey background overlaid with more important critical information such as a live electrical circuit (Amber) and a running Gas Turbine (Green). Note that the colours vary widely within and across different military systems due to the difficulty in aligning all systems and industries to a common colour standard, together with important variations in operational requirements. Show the computer RGB (Red, Green, Blue) values to facilitate faithful reproduction at all times, across all sub-systems (Table 5.1).

Amount of Colour – Use colours sparingly leaving plenty of space around objects to avoid clutter.

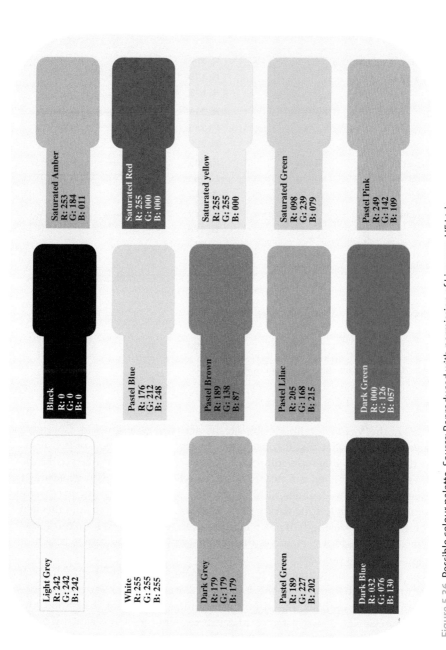

Saturated Amber
R: 253
G: 184
B: 011

Saturated Red
R: 255
G: 000
B: 000

Saturated yellow
R: 255
G: 255
B: 000

Saturated Green
R: 098
G: 239
B: 079

Pastel Pink
R: 249
G: 142
B: 109

Black
R: 0
G: 0
B: 0

Pastel Blue
R: 176
G: 212
B: 248

Pastel Brown
R: 189
G: 138
B: 87

Pastel Lilac
R: 205
G: 168
B: 215

Dark Green
R: 000
G: 126
B: 057

Light Grey
R: 242
G: 242
B: 242

White
R: 255
G: 255
B: 255

Dark Grey
R: 179
G: 179
B: 179

Pastel Green
R: 189
G: 227
B: 202

Dark Blue
R: 032
G: 076
B: 130

Figure 5.36 Possible colour palette. *Source:* Reproduced with permission of Liveware HF Ltd.

Table 5.1 Example recommended pastel colours for RN IPMS pipe in-fills.

System Items	Colour	RGB Value	Example
No Flow/Current (in-active)	**Grey**	**R: 179** **G: 179** **B: 179**	
Air	**White**	**R: 255** **G: 255** **B: 255**	
Ship's Dieso (Diesel)	**Lilac**	**R: 205** **G: 168** **B: 215**	
Lub Oil	**Pastel Brown**	**R: 189** **G: 138** **B: 87**	
Chilled Water	**Pastel Blue**	**R: 176** **G: 212** **B: 248**	
HPSW	**Pastel Green**	**R: 189** **G: 227** **B: 202**	

Source: Reproduced with permission of Liveware HF Ltd.

5.9.3 Colour Perception

In 1941, the vision scientist Selig Hecht and his colleagues at Columbia University made what is still considered a reliable measurement of the 'absolute threshold' of vision - the minimum number of photons that must strike the retina in order to elicit an awareness of visual perception. The experiment probed the threshold under ideal conditions: study participants' eyes were given time to adapt to total darkness, the flash of light acting as a stimulus had a (blue-green) wavelength of 555 nanometers, to which the eyes are most sensitive, and this light was aimed at the periphery of the retina, which is richest in light-detecting rod cells.

The scientists found that for study participants to perceive such a flash of light more than half the time, the subjects required between 54 and 148 photons to hit their eyeballs. Based on measurements of retinal absorption, the scientists calculated that a factor of ten fewer photons were actually being absorbed by the participant's rod cells. Thus, the absorption of five to 15 photons, or, equivalently, the activation of just 5 to 15 rod cells, tells the brain that something is seen.

Coloured objects and easily readable text need to subtend 0.5% of the viewing distance, i.e. about 3 mm at 600 mm across the smaller dimension.

5.10   Text

The supplier should provide a description of all of the text used as part of the system HCI and deviate only with prior agreement. Consistency should be maintained wherever possible. Choice of text size is critical to human performance and research conducted for DERA by Liveware in 1997 is worthy of scrutiny. The curve in Figure 5.37 below is notably steep on both sides suggesting that users seem to prefer a definite and optimum character size, in this case about 20 arcmin (angle subtended at the eye) high (and 18 arcmin wide). As a rule of thumb, character height should be about 1% of the viewing distance and should be neither too small nor too large:

5.10.1   Font Type

The Arial (san serif) font is recommended throughout the system HCI.

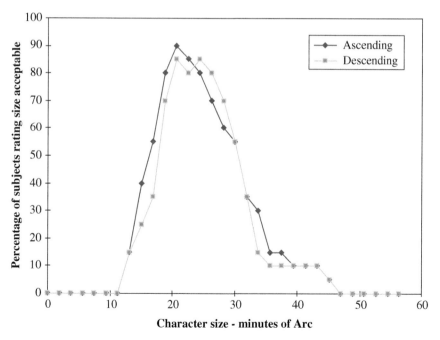

Figure 5.37   Preferred character height from subject trials. *Source*: Reproduced with permission of Liveware HF Ltd.

Table 5.2 below specifies the characteristics of the various text elements of the system HCI.

Table 5.2 Example text characteristics table.

Text Element	Size	Case	Colour	Weight	Just'	Example
Page Title	24 pt	Mixed	Black	Plain	Centre	**Not provided**
Navigation Group Button Label	11 pt	Upper	Black	Bold	Centre	
System Navigation Button Label	11 pt	Mixed	Black	Bold	Centre	
Hyperlink Button (large) Label	12 pt	Mixed	Black	Bold	Centre	
Hyperlink Button (medium) Label	11 pt	Mixed	Black	Bold	Centre	
Hyperlink Button (small) Label	8 pt	Mixed	Black	Bold	Centre	
Dialogue Box - Variable	12 pt	Mixed	Black	Bold	Varies	
Dialogue Box – Running Status (Large)	16 pt	Mixed	Black	Bold	Centre	
Dialogue Box – Running Status (Small)	12 pt	Mixed	Black	Bold	Centre	
Pop-Up Window Title	10 pt	Mixed	Black	Bold	Centre	
Start and Stop Controls (Disabled)	14 pt	Mixed	Grey R121 G121 B121	Bold	Centre	

Source: Reproduced with permission of Liveware HF Ltd.

5.11   Symbology

The supplier should provide a description of all symbols and icons used as part of the system HCI and deviate only with prior agreement with the Authority. Consistency should be maintained wherever possible.

5.11.1   Marine/Systems Engineering

Table 5.3 below details example marine/systems engineering icons/symbols to be provided as part of the system HCI.

Icons should change colour depending upon their running status (e.g. the DG icon) or their position (e.g. a controllable valve).

Table 5.3 **Example symbols and icons table.**

System Element	Icon/Symbol
DG	
Shaft	
Steering Pump	
Gearbox	
Diesel	
Gas Turbine	

(Continued)

Table 5.3 (Continued)

System Element	Icon/Symbol
Generator	
Transformer	
Breaker (Horizontal)	
Breaker (Vertical)	
Non–Return Valve (One Way)	
Non-Return Valve (Two Ways)	
Non-Return Valve (Three Ways)	
Pump	
Valve (Local Control)	
Valve (Remote Control)	
Cable	
SWBD	Emergency LV SWBD
Fluid/Air Pipe	

Source: Reproduced with permission of Liveware HF Ltd.

Where status information is available (e.g. a DG will always be provided with a Running Status Dialogue box), the icon should change colour to reflect the alert state.

Some icons should flash to indicate when the equipment is in a transitional phase. The coding for the DG symbol should be used and is provided as an example below (Table 5.4):

Table 5.4 Example of DG icon presentation states.

Icon Presentation State	Illustration
Available	
Starting	
	Flashes between the above 2 states (1 complete cycle every 1.5 seconds)
Running	
Synchronising	
	Flashes between the above 2 states (1 complete cycle every 0.8 seconds)
Shutting Down	
	Flashes between the above 2 states (1 complete cycle every 1.5 seconds)

(*Continued*)

Table 5.4 (Continued)

Icon Presentation State	Illustration
Stopped	
Not Available	
OOA	
In Warning	
In Alarm	
System Connection Lost	

Source: Reproduced with permission of Liveware HF Ltd.

5.12 Mimics

5.12.1 General

The supplier should provide details of mimic pages to be provided in terms of operation, look and feel, and deviate only with prior written agreement. Issues to be covered where relevant include:

1) Orientation;
2) Prioritisation of information;

3) Layout;
4) Grouping of components;
5) Labelling;
6) Zone boundaries;
7) Ship/compartment outlines;
8) Dynamic piping.

The following sub-sections provide examples of different types of mimic page/components and associated design features.

5.12.2 Ringmain Mimics

The following figure provides an example of a marine HPSW ringmain mimic display page with the associated design features (Figure 5.38):

1) Orientated forward to the right, aft to the left, Port at the top, Starboard at the bottom;
2) Control panels for pumps provided with background panels and located within the ringmain;
3) Variable dialogue boxes located on the pipes;
4) Identity of valves available by the operator hovering the pointer over the valve for two seconds;
5) Dynamic piping provided to re-affirm the position of valves and pump running status;
6) Zone labels provided in 12 pt bold text;
7) Aft, Forward labels provided in 18 pt bold text;
8) Zone boundaries provided as dark grey dashed lines (R:83, G:83, B:83), with a weight of four pixels;
9) Small hyperlink buttons integrated with piping (Figure 5.38).

5.12.3 Electrical Mimics

The following figure provides an example of a marine electrical mimic display page with the associated design features that should be implemented (Figure 5.39):

1) Generally orientated with forward at the top of the page, aft to the bottom, Port to the left and Starboard to the right. (However, it is considered more important by RFA (Royal Fleet Auxiliary) operators for the flow of power from HV to LV to be from top to bottom. As a result, in the example at Figure 5.42, the Port aft LV SWBD is positioned above the Port forward LV SWBD and the Starboard aft LV SWBD is positioned above the Starboard forward LV SWBD. The orientation of these elements is maintained throughout all display pages to ensure consistency).

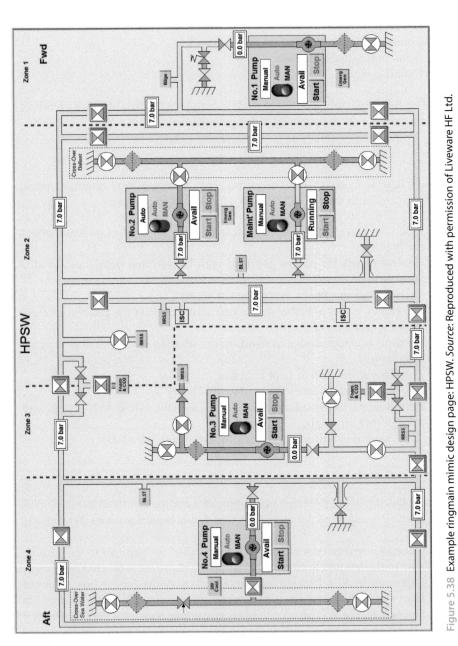

Figure 5.38 Example ringmain mimic design page: HPSW. *Source*: Reproduced with permission of Liveware HF Ltd.

2) Equipments supplying the switchboards should be positioned above the busbars and equipments being fed by the switchboards should be positioned below the busbars.
3) Background panels to group associated information.
4) Variable dialogue boxes located on the cables where possible.
5) Identity of breakers available by operator hovering pointer over valve for two seconds ('Tooltip').
6) Dynamic cabling provided to re-affirm position of breakers and running status of power supplies.
7) Hyperlinks should be integrated into the mimic design.
8) Differentiation of HV from LV by use of colour and thickness of busbars (Figure 5.39).

5.12.4 Propulsion Mimics

The following figure shows an example of a propulsion mimic display page where the following associated design features should be provided (Figure 5.40):

1) Orientated forward to the top, aft to the bottom, Port to the left and Starboard to the right;
2) Background panels to group associated information;
3) Dynamic cabling provided to reaffirm position of breakers and running status of power supplies;
4) Identity of breakers available by the operator hovering the pointer over the valve for two seconds;
5) Equipments supplying the switchboards are positioned above the busbars and equipments being fed by the switchboards are positioned below the busbars;
6) Differentiation of HV from LV by use of colour and thickness of busbars;
7) Where possible hyperlinks are integrated into the mimic design (Figure 5.40).

5.12.5 Tank Gauges

The following figure shows an example of a tank gauge where the associated design features should be provided (Figure 5.41):

1) Label provided at top (Black, 11 pt Bold);
2) Combination of percentage contents provided as a mimic and cubic meters provided as a variable dialogue box so that contents can be ascertained 'at-a-glance' yet an accurate reading recorded during inspection rounds if required;
3) Fluid colour re-affirms the system the tank related to (Figure 5.41);

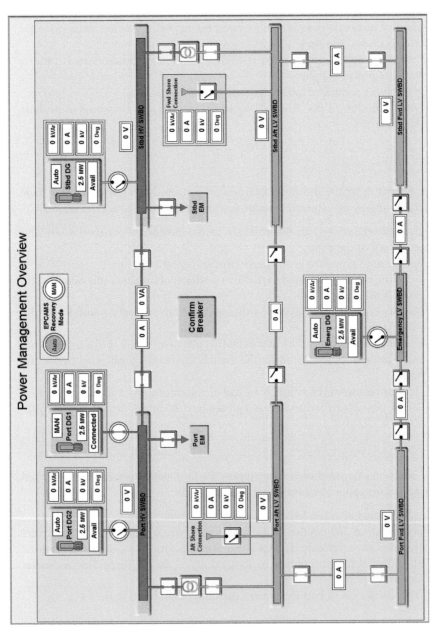

Figure 5.39 Example electrical mimic design page: Power management overview. *Source:* Reproduced with permission of Liveware HF Ltd.

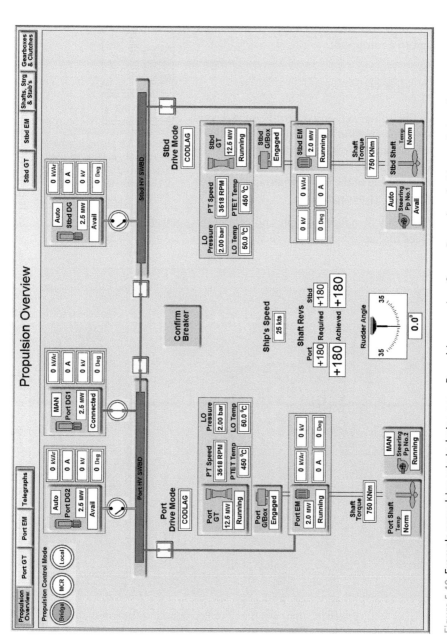

Figure 5.40 Example propulsion mimic design page: Propulsion overview. *Source:* Reproduced with permission of Liveware HF Ltd.

5.12.6 Animation

The supplier should describe any animation to be provided and deviate only with prior agreement. Repetitive animation can have similar problems to excessive use of colour and should be kept to a minimum.

5.13 Touch Screens

Generally, touch screen control is only recommended for specific applications that enjoy a significant benefit from touch screen technology (Figures 5.42, 5.43). Mobile phones and iPads are a good example of the appropriate application of this technology because:

Sump tank

% Contents
— 100
— 80
— 60
— 40
— 20
— 0

553 m³

Figure 5.41 Example tank gauge: Lub oil tank. *Source:* Reproduced with permission of Liveware HF Ltd.

1) Portability demands a simple and lightweight solution;
2) Speed and accuracy is not critical;

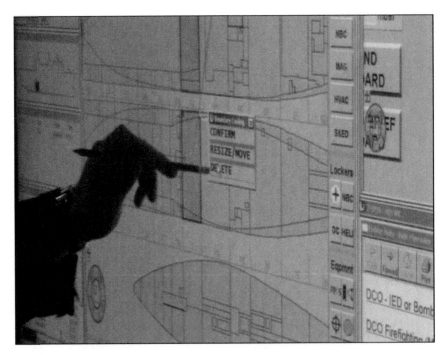

Figure 5.42 Touch screen HCI using stylus. *Source:* Reproduced with permission of Liveware HF Ltd.

Figure 5.43 Touch screen HCI using finger.

3) Errors have little consequence;
4) The touch input can be invoked only when needed, thus freeing up maximum glass area;
5) They are immediately intuitive to the user, require minimal training and therefore promote rapid usability on the move.

The following issues are critical to the success of touch screen technology as an HCI such that careful thought needs to be applied when used in terms of the design of the HCI:

1) Not recommended for moving platforms whereby the platform motion could disturb the speed and accuracy of the input.
2) Not recommended where speed and accuracy is critical.
3) Not recommended for critical inputs where safety is paramount.
4) Where gloved hands are needed, the touch screen target area should be at least 30 mm square. This could be achieved by temporarily increasing the size of the target area when hovering over it with a finger or stylus.
5) Where gloved hands are needed, the touch screen target area should respond to the glove material hitting the screen; not the case with some touch screen technologies.

6) For standing operators against a vertical screen the height of the data entry point is critical for performance and user comfort: chest height is best at about 1500 mm above the ground. Horizontal or sloping touch screen surfaces are easier to use because the user can support the wrist for greater comfort and precision but the wrist must not invoke an input.

7) Sometimes a stylus input device – the rubber end of an ordinary pencil – can enhance usability and gets over the gloved hand problem provided the stylus itself can activate the screen surface.

8) Good for 'hunt and peck' entries but not extensive 'narrative' entry.

9) Fatiguing when used for more than a few minutes – an arm rest is essential for prolonged use.

10) Touch screens necessitate the user being close to the screen thus restricting the view from other users.

11) Touch screens require a specific HCI design to meet user requirements.

12) Sometimes it might be prudent to combine a touch screen interface with the capability to use other input devices as a back-up or alternative, for example, a trackball.

13) Constant use of a touch screen with dirty or greasy hands will obscure the display, requiring regular cleaning.

14) Touch screens need additional user re-enforcement to show first, that the area of the screen is available for data entry, secondly, the activation has been accepted by the system and thirdly that an effect has occurred such as the opening of a valve, closing of a breaker or other such action. Tactile feedback could be augmented by a vibration, bleep or visual change in the surface appearance of the soft button area.

15) For critical systems it could be beneficial to have a dedicated hard button to lock the touch screen to avoid accidental activation.

In summary, touch screens may be advocated for equipment occasionally used but are not generally liked for primary HCIs on a moving platform under high workload, safety critical conditions.

5.14 Day and Night Viewing Conditions

5.14.1 Night Viewing Palettes

The Supplier shall provide a description of all of the colours provided as part of the HCI when it is in Night Viewing mode. An example day/night mimic is provided below and it is recommended that a dedicated control is provided in order to invoke day or night mode with a single control action (Figures 5.44, 5.45):

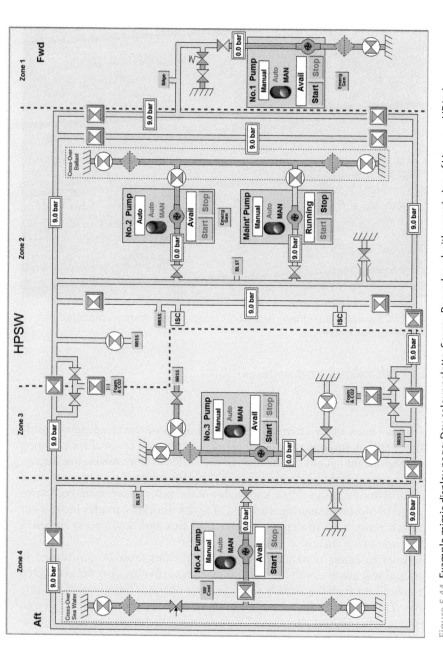

Figure 5.44 Example mimic display page: Day colour palette. *Source:* Reproduced with permission of Liveware HF Ltd.

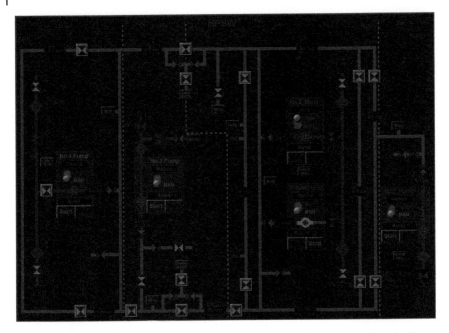

Figure 5.45 Example mimic display page: Night colour palette. *Source*: Reproduced with permission of Liveware HF Ltd.

5.15 Workload and Automation

5.15.1 Workload

There have been many attempts to measure operator workload, some overly complex and of dubious validity. There is no substitute as usual for common sense. Significant insight can be derived on workload from qualitative inspection of operator tasks performed within a specific scripted scenario, provided that the observer has in-depth knowledge of the tasks undertaken. For example, any HF person measuring workload of a pilot during a complex instrument approach must have in-depth experience of that task and access to SMEs (pilots) whom are current at that task.

In order to assess workload, the critical activities must be scrutinised in detail and will involve both human and machine activities. Any data collected will have a long shelf life, usually years, because operator tasks tend to be fairly stable. This data can be re-visited downstream for retrofits or re-design purposes with minimal alteration. There are many techniques to assess operator workload and the best are the simplest, focusing on face value observation of obvious bottlenecks arising from complex tasks. The value of workload assessments is that control rooms can be better designed if the reasons for high workload are understood. Usually, high workload will arise from one, or a combination of, the following factors, based fundamentally on the task, operator, equipment (HMI/HCI design) and environmental conditions:

1) Poorly designed processes or task procedures, e.g. poor checklists or operating procedures;
2) Poorly trained operators, e.g. receiving no training or the wrong training;
3) Operators not suited or qualified for the job, e.g. poor selection procedures;
4) Poorly designed equipment, e.g. HMIs/HCIs;
5) Demanding environmental conditions, e.g. poor flying visibility, excessive heat/cold, noise, tiredness, etc.

The following figure (Figure 5.46) provides an example workload chart based on a timeline of Events (horizontal scale) against the Operators Involved (vertical scale) showing levels of workload intensity. This example is taken from workload assessments of aircraft carrier operations room talk-down controllers to determine how many were required. The figure is truncated at the right hand side since the original continues for many more minutes. The diagram is explained as follows:

1) **Events** – key activity milestones as part of a scripted scenario;
2) **Workload Intensity** – single operator actions undertaken within a specified time;
3) **Operators Involved** – optimum manning to share the tasks.

In summary, workload assessment requires identification of the key functions of the system which need to be decomposed down to activity level then shared optimally amongst the candidate staff as shown (Figure 5.47).

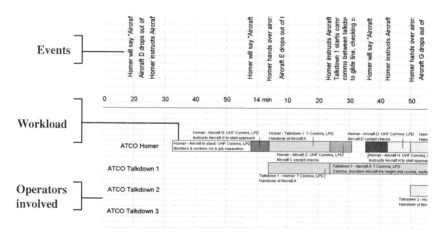

Figure 5.46 Example workload chart. *Source*: Reproduced with permission of Liveware HF Ltd.

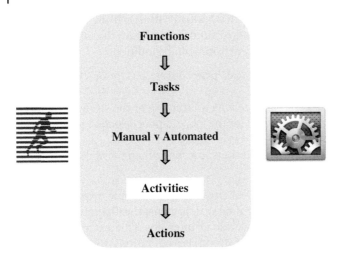

Figure 5.47 Functional breakdown. *Source*: Reproduced with permission of Liveware HF Ltd.

Automation needs careful consideration, notably in providing the user with feedback at all times as to what the automated equipment is doing.

5.15.2 Automation and De-skilling

Since WWII, technology has advanced to a stage whereby a high proportion of human tasks have been replaced by computer-based automation. Automation has numerous advantages over manual human control including the ability to undertake dangerous tasks, tasks which are very monotonous, tasks that can be undertaken more efficiently by machines and tasks that are beyond human capability. Examples include vehicle ABS braking, aircraft autopilots, automatic plant control in an oil refinery, automated vehicle production lines, satellite navigation and, more recently, driverless cars.

It follows that automation can, in the main, be a good thing and is here to stay, provided that it is managed appropriately through an understanding of human ability and behaviour. To resist automation is not to reject it. The following paragraph takes brief quotes from Nicholas Carr's enchanting book *The Glass Cage* (Ref. 5). It is so transfixing that it has a place in this HF Guide to support the importance of good control room design; it should be on the reading list of all equipment designers.

5.15.2.1 Anecdotal Evidence

GPS Navigation - Carr's book narrates a story of Inuit hunters in the Canadian North who have, for some four thousand years, ventured from their homes on the island of Igloolik and traversed miles of ice and tundra in search of food.

The hunters' ability to navigate vast stretches of barren Arctic terrain, where land-marks are few, snow formations are in constant flux and trails disappear overnight 'has amazed voyagers and scientists for years'. Carr describes how they have done this without maps or compasses, but with a profound under-standing and perception of winds, snowdrift patterns, animal behaviour, stars, tides and currents. Around the year 2000 the Igloolik hunters, especially the younger ones, began to rely on computer generated maps and GPS to get around. By purchasing a cheap GPS the life-long arduous training could be skipped and journeys could be accomplished at any time, e.g. in dense fog. Carr then points out that 'as GPS devices proliferated, reports began to emerge of serious accidents during hunts, of injury and even death'. When the technology broke down, e.g. loss of satellites, loss of signal or flat batteries, a hunter not conversant with the ancient skills easily became lost in the featureless waste and fell victim to starvation and exposure. Even when operating correctly, the modern devices 'can give hunters a form of tunnel vision, following the elec-tronically created routes faithfully onto thin ice, cliff edges or other environ-mental hazards which, hitherto, their instincts would have warned them against'. It is these very instincts, wisdom, knowledge and experience that auto-mation can render faded and un-used in humans. The Inuits were starting to travel 'blindfolded'.

Boeing vs Airbus Cockpit Design - The issues of technology-centred ver-sus human-centred automation are critical in order to achieve a balanced auto-mation philosophy. The two dominant airliner manufacturers have approached the design of flying controls in cockpit design in markedly different ways, both based on fly-by-wire systems. Airbus has pursued a more technology-centred design in deploying small side-mounted joysticks to receive commands from the pilot to the flying control surfaces. The Captain uses his left hand, the Co-pilot his right hand. This raises separate issues should one have to swap places with the other. These small game-like joystick movements send inputs to the flight computer which can intervene and override pilot inputs if pre-specified flight profile boundaries are violated. Further, they can invoke pre-programmed profiles that ensure the most efficient configurations of control surfaces and power to optimise flight efficiency. In theory, this stops pilots putting the air-craft into configurations which could be near to or outside of the safe flying envelope. There is little tactile feedback (unlike a conventional flying control) and this design emphasises the Pilot's role as a computer operator rather than an aviator. Further, it is not easy for the non-flying co-pilot to monitor (see) the instantaneous position of the joystick, nor the intentions of that pilot. The soft-ware, not the pilot, wields ultimate control. Boeing has taken a more tradi-tional, human-centred approach. The aviator in a Boeing aircraft retains final authority over manoeuvres, even in extreme circumstances. Large, centrally mounted yokes are deployed for use by one or two hands if necessary and they follow, faithfully, the direct inputs from the pilot. Further, although power assisted, they are loaded ('control loading') with resistance in order to provide measured tactile feedback so that the pilot always knows how much effort (force) is being applied and precisely how far (linear distance) that movement has been executed in any direction for pitch and role. They feel like flying a real plane. Imagine driving a car with a computer joystick and how that would feel, and how you might react if the car got into a skid and the automated assistance failed. Research has shown that tactile (haptic) feedback is significantly more effective than visual cues alone in alerting pilots to important changes in a plane's orientation (Ref. 6). Further, because the brain processes tactile signals in a different way from visual signals, haptic warnings don't tend to interfere with the performance of concurrent visual tasks. Thus Boeing pilots wielding large yokes are more involved in the bodily experience of flight than their counterparts on an Airbus manipulating small joysticks. Since both Airbus and Boeing aircraft have phenomenal safety records it is likely that there are advantages with each design, if only they could be combined to enjoy the synergy of these benefits.

The purpose of this HF Guide is to bring the various HF issues to the attention of control room designers so that they can investigate and debate the relative merits of the different approaches and choose what is best to create balanced automation. Some aviation experts believe that the design of the Airbus cockpit played a part in the Air France Flight 447 disaster on 1 June 2009. The BEA's (Bureau d'Enquêtes et d'Analyses pour la Sécurité de l'Aviation Civile) final report, released at a news conference on 5 July 2012, concluded that the aircraft crashed after temporary inconsistencies between the airspeed measurements, likely due to the aircraft's pitot tubes being obstructed by ice crystals. This caused the autopilot to disconnect, after which the crew reacted incorrectly and ultimately led the aircraft to an aerodynamic stall from which they did not recover. The pilot flying the aircraft was pulling back on his joystick whilst his co-pilot was oblivious to this fateful error, causing the aircraft to stall. In a Boeing cockpit, each pilot has a clear view of the other pilot's yoke and how it is being manipulated. Further, the two yokes operate as a single unit; if one pilot pulls back on his yoke, the other pilot's goes back too. Through both visual and tactile cues, the pilots are tuned into each other's actions and intentions. The Airbus joysticks are not so much in clear view and operate independently, with more subtle movements. Any stress induced, typical of situations requiring manual override, would further amplify any design deficiencies and decreased human awareness. It is possibly unfortunate that some reports on similar accidents point towards operator error being the main cause when, in fact, the cause has more to do with poor design of the HMIs/HCIs, over reliance on automation and lack of practice in manual intervention. It is not the purpose of this HF Guide to apportion blame nor to criticize any specific company's products but particular emphasis in this Guide is placed on attention to scrutiny of Workspace and HMI/HCI design.

5.15.2.2 Balancing Automation between the Human and the Machine

Early navigation aids such as maps and compasses gave travellers a greater awareness of their environment, sharpening, not blunting their sense of direction. However, much automation is not designed to deepen our understanding and involvement of our Situational Awareness (SA); it replaces it and relieves the user of effort, in effect, de-skilling the task. With a modern GPS, users no longer need SA to know where they are and, spatially, where their destination is, or attend to natural waypoints along the way. Everything is done for them so the task of navigation is no longer immersive or involving and their natural instincts are inhibited. Nicholas Carr's book *The Glass Cage* (5) goes on to say 'If humans never have to worry about not knowing where they are, then they never have to know where they are'. Automated navigation enables users to get around without needing to know where they are, where they've been, nor in which direction they're heading. The modern brain is no longer perceiving the raw material (natural data) required to form rich memories of places and space. The act of manual navigation forms strong bonds with our cognitive (thinking) and psychomotor (muscle control) skills; Carr (5) says 'where we are is who we are' and is tied into our innate awareness of our surroundings. Traditional maps give us context, provide SA and scale and act as cognitive levers to sit alongside our own innate skills. Further, they enable us to enjoy the journey, not just the destination. Dis-orientation, for example, acts as a catalyst, pushing us to a fuller awareness and deeper understanding of our situation. Experiments have shown that people's navigational senses are sharpest when they're facing north, presumably because, from an early age, most of us have looked at maps which are oriented north up. Using maps teaches us to think about space. For a similar reason, operators in moving platform control rooms should, where feasible and appropriate, face in the principal direction of motion of the platform, usually forwards.

A logical advance of GPS is indoor mapping, e.g. using Google Glasses as head-up displays to provide a continuous stream of guidance around buildings. Could long term use of these devices weaken our memories and senses and raise the risk of premature senility? Not if we design counter-balancing strategies to ensure that technology and humans work in harmony as human-machine systems. Carr (5) goes on to point out that 'This argument can be extended to the proliferation of computer aided design (CAD) as essential tools for architects and designers. They now not only turn designs into plans, they produce the designs themselves. However, these machines are in danger of replacing the designer's untidy and painstaking process of exploration, thought, creativity and innovation'. Good architecture, for example, is about taking time, making mistakes and re-iteration through a comet trail of manually involved ideas. Like a true craftsman, the accomplished architect takes a design and breaks it down into parts that need the attention to detail of a jeweller or cabinet maker. Drawing with a pencil is also an act of thinking; a conduit between the hand and the brain. As ideas take form a creative spark ignites and begins its slow migration from imagination to design. The physical act of drawing, mental effort and deliberate muscle movements, form long-term memories and a bonding with the true intent of the design. Sketching enables exploration of the problem space and the solution space to proceed together, unlocking the mind's hidden stores of tacit knowledge.

For the interested reader, outside of the scope of this book, navigation is understood to be undertaken using two key areas of the brain. The hippocampus, concerned with 'place cells' in remembering waypoints and the entorhinal cortex, concerned with 'grid cells' in the appreciation of relative space in an array of regularly spaced equilateral triangles. These neural mechanisms that evolve to define the spatial relationship among landmarks can also serve to embody associations among objects, events and other types of factual information. Out of such associations the memories of our lives are weaved. It may well be that the brain's navigational sense (its ancient, intricate way of plotting and recording movement through space) is the evolutionary font of all memory. One of the earliest and most debilitating symptoms of dementia, including Alzheimer's disease, is hippocampal and entorhinal degeneration and the consequent loss of locational memory.

It follows that automation must work with users not just for them. Tasks must not only be appropriately allocated between humans and machines, they must be designed to be undertaken in a collaborative way so that the human doesn't 'tune out'. Humans have 'tacit' knowledge that is learned from experience but is difficult to explain in detail. For example, a person can be shown how to fly-fish but not made to acquire the complete skill intuitively. Machines have 'explicit' knowledge programmed into them that can be written down or scripted. For example, a machine can obey a set of automated procedures that apply only within a small sphere of understanding. Autonomous machines will eventually come across moral dilemmas, e.g. with a self-driving car they will have to make decisions; they must be programmed eventually, to make ethical decisions. Whom does the programming? Can a machine be ethical? To whom is a court writ served in the event of an accident? A robotic car would not jump a red light where a human might, but it will not cope with a situation it isn't programmed to cope with. That suggests that automation should be retained for simple procedural tasks. Humans are much better at coping with complex and novel situations, e.g. Apollo 13, where experience is required beyond the bounds of a precise, bounded mission.

Control room operators require an optimal level of stimulation, not too under-worked and not too over-worked. Automation in control room design must be a contract between the machine and the operator that they work together, not distancing humans from the natural environment that originally shaped their behaviour. Instead of encouraging the manipulation of symbols on a computer screen it should promote our attention to real things and real places, bracing us to 'be here now'. Automation is in danger of drawing the operator into a synthetic, seemingly unreal world, not unlike a computer game.

5.15.2.3 Key HF Issues in Addressing Automation

Addressing automation requires getting to grips, intellectually, with the key operator tasks. For an HF expert to undertake any kind of useful design requires experience and an in-depth understanding of these tasks; not expertise in documentation, over-analysis, excessive computer modelling or recording laborious data on spread-sheets, etc. Sleeves have to be rolled up and a small risk taken in 'doing the job' for which the design is intended. HF experts must fly the plane, operate the refinery as a supervisor, manage a damage control exercise on a warship; get inside the 'mind set' of the operator. In summary, taking account of the complex issues raised in the above paragraphs, in HF terms, some of the fundamental rules in adopting automation are that automated systems should:

1) Be regularly monitored by humans for any dangerous deviation or failure.
2) Be able to be over-ridden quickly by humans in the event of failure.
3) Provide constant feedback to the human (and colleagues) on their status in a clear and unambiguous way. For example, manual flying controls on some aircraft should clearly show their exact positions and rates of change when the autopilot is engaged, providing direct and unambiguous indications of how the aircraft is being flown. Thus if intervention is necessary in an emergency, there is no doubt as to the immediate control actions required.
4) Always keep the human in the loop, and in overall total control, as to system status so that automation is still part of a human-machine system (Figure 5.48).

Figure 5.48 Balanced human-machine interaction with the human always in overall control. *Source*: Reproduced with permission of Liveware HF Ltd.

5) Not de-skill the human whereby the retention of critical skills is jeopardised. For example, these vital skills might include the ability of a pilot to fly a plane manually, immediately after autopilot failure in stressful conditions or the ability of a hiker to retain basic navigational skills, using the natural surroundings, following GPS malfunction or complete failure, in poor weather. This de-skilling is not a direct function of automation, it is a training and procedural issue to ensure skill retention and needs to be managed.

6) Be driven by logical and related processes based closely on existing, previous manual procedures. Operators using newly automated systems are likely to have been previously trained on more basic and proven system to which they must be able to relate. This is especially important if the automation fails and the operator has to quickly (often under duress) revert to back-up systems. Under these circumstances, humans usually resort back to pre-learned procedures because these are indelibly ingrained through previous training and experience.

7) Include hard copy manuals that might need to be referred to in the event of system failure or when system operation requires refreshed learning. Computer based instruction manuals are no use if the computer systems have failed; on-screen manuals can block out the very information the user is trying to find out about. For example, it might be prudent to carry manual navigation charts for the area of operation on a fully automated ship's bridge. Quick reference hard copy user guides could be useful for parallel access in using the electronic screens. Ensure that space is available within the workplace to use manual back-ups (e.g. a ledge under a Bridge window to rest the elbows if having to use binoculars) and any manual instruments (e.g. a hand-bearing compass, pencils, rubbers, dividers, parallel rulers, etc.). Usually, providing plenty of uncluttered horizontal surfaces at the correct height (seated or standing) is all that is needed.

8) Keep the operator at an optimum level of workload and involvement in the tasks; not too idle and not too busy. This could be achieved by ensuring that the human intervenes at regular intervals, e.g. flies the plane, drives the train, opens a few processing valves in a refinery plant, plots a fire boundary around a damaged warship compartment, takes a bearing on a passing ship, etc..

9) Not leave individuals unaccompanied in control rooms; both the human and the machine require supervision.

6

Environmental Ergonomics

6.1 Outline

This section covers those factors relevant to human performance, comfort and safety that arise from the surrounding environment. They are therefore not directly related to equipment design but should be taken into account by Suppliers so that their equipment can be optimised for the workspace where they will be deployed. Environmental ergonomics includes:

1) lighting;
2) noise and vibration;
3) heating and ventilation;
4) platform motion.

6.2 Lighting

In control room design, lighting is concerned with two key workspaces:

1) Ambient lighting around workplaces;
2) Task lighting within workplaces.

Ambient Lighting - is generally set at a lower level than task lighting because it is provided for movement around control rooms or passageways without disturbing other areas.

Task Lighting - is generally set at a higher level than ambient lighting because it is provided for tasks within a workspace, sometimes requiring very high levels for intricate work.

VDU-based Control Rooms - Where tasks involve the use of VDUs which, themselves, are self illuminated, too much ambient and task lighting can wash-out screen information. For these control rooms, ambient lighting is best placed high up and directed upwards, to reflect off a light matt ceiling or

Human Factors in Control Room Design: A Practical Guide for Project Managers and Senior Engineers, First Edition. Tex Crampin.
© 2017 John Wiley & Sons Ltd. Published 2017 by John Wiley & Sons Ltd.

Table 6.1 Approximate ambient and task lighting values (lux).

Location	Lighting Level (lux)
Compartments rarely used - movement access	ambient 50
Compartments rarely used - basic object detail	ambient 100
Compartments rarely used - hazardous	ambient 150
Compartments used for long periods - basic object detail	ambient 200
Control Rooms - office tasks with VDUs	ambient up-lighting 500
Control Rooms - detailed inspection tasks	Task lighting 1,000
Control Rooms - precision inspection tasks	Task lighting 3,000

Source: Reproduced with permission of Liveware HF Ltd.

Table 6.2 Control room surface reflectance (%).

Surface (non gloss)	Reflectance (%)
Floor	30–40
Walls	50
Ceiling	80

Source: Reproduced with permission of Liveware HF Ltd.

deckhead. Task lighting (Table 6.1) needs to be especially directional and not pointing directly at the VDUs.

Reflectance - Compartments should generally provide the following reflectance (%) values of walls, floors and ceilings (Table 6.2).

6.3 Noise

Noise is defined as unwanted sound and too little or too much can affect human performance in terms of the ambient working noise levels.

The A-weighting is an electronic weighting network with a frequency response which approximates to that of the human ear by de-emphasising low frequencies and extreme high frequencies. Sound levels measured with the A-weighting in a sound level meter are referred to as A-weighted sound levels and quoted in dB(A).

Def Stan 00-250 Section 14 quotes an 8 hr working day period limit of 80 dB to 85 dB. Peak noise limits range from 135 dB to 137 dB. Note that since the dB scale is logarithmic, a 3 dB increase in sound power amounts to an approximate doubling of the sound level power.

Maximum noise levels are recommended in Table 6.3 below which, although for ships, should apply approximately the same to other industries.

For hearing conservation purposes, continuous noise (from machinery, vehicles, aircraft, etc.) is measured either in terms of the A-weighted sound level expressed in dB(A), or in terms of the equivalent continuous sound level (8 hour), usually written as Leq (8 hour) and also expressed in dB(A). The limit of noise exposure reported by the Defence Standard 00-250 is an 8-hour Leq at the ear of 85 dB(A); this Leq shall not be exceeded during any 24-hour period. It is, however, strongly recommended that, as far as possible, the sound level at the ear should not exceed 85 dB(A). The relation between Sound Level and Duration is summarised in the following Table 6.4 for an Leq of 85 dB(A).

Table 6.3 Approximate recommended noise levels: IMO ships.

Compartment	Max (dB(A))
M/C Space continuous occupancy	90
M/C Space occasional occupancy	110
M/C Control Room	75
Ship's Bridge	65
Accommodation Cabins	60
Galley	75

Source: Reproduced with permission of Liveware HF Ltd.

Table 6.4 Relation between sound level and duration for an Leq of 85 dB(A).

Sound level	Maximum Continuous Exposure
80 dB(A)	24 hours
82 dB(A)	16 hours
85 dB(A)	8 hours
86 dB(A)	6 hours
88 d(B(A)	4 hours
91 dB(A)	2 hours
94 dB(A)	1 hour
97 dB(A)	30 minutes
100 dB(A)	15 minutes

Source: Reproduced with permission of Liveware HF Ltd.

Table 6.5 Human responses to a range of effective temperatures.

°F	°C	Responses
110	43	Just tolerable for brief periods. Human tissue begins to burn.
90	32	Upper limit of tolerance, becoming painful.
80	26	Extremely fatiguing to work in. Performance deteriorates badly and people complain a lot.
78	25	Optimal for bathing, showering. Sleep is disturbed.
75	24	People feel warm, lethargic and sleepy. Optimal for unclothed people.
72	22	Most comfortable year-round indoor temperature for sedentary people.
70	21	Optimum for performance of mental work.

Source: Reproduced with permission of Liveware HF Ltd.

6.4 Heating and Ventilation

A generic way to describe the thermal environment is by means of what is called the 'effective temperature'. It is important to note that there are several effective temperature scales, the most useful being designated ET (Table 6.5). Effective temperature is equal to air temperature when the relative humidity is 50%, the air movement is 0.1 metres per second and when the walls of the room are at the same temperature as the air.

Air flow is also important, especially through control rooms containing several operators. Air ducts should not be directed onto individuals and should be introduced at 0.007 m^3/sec/person. Air should be circulated at no more than 0.25 m/sec. Modern warships use text venting consisting of a large sock (at the end of a duct) whose tiny holes allow air to permeate slowly and evenly without causing draughts.

Figure 6.1 below shows a graph of relative humidity (R/H) (%) against temperature (°C). The comfort zone is neither too hot nor too cold and neither too damp nor too dry.

6.5 Platform Motion

The motions of military vehicles and platforms can have adverse effects on the comfort, well-being and task performance of personnel. In some circumstances, health and safety may also be compromised. There are four primary phenomena:

1) **Motion sickness** – caused by low-frequency motion occurring with both short and long-term exposure. Sickness occurs as a result of exposure to

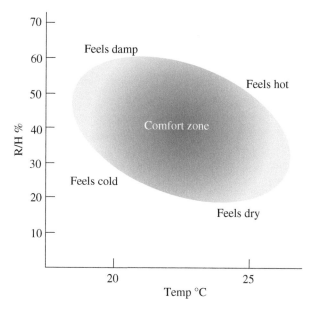

Figure 6.1 Relative humidity (R/H) vs. temperature (°C). *Source:* Reproduced with permission of Liveware HF Ltd.

mainly vertical (heave) motion in the frequency range 0.05 to 0.7 Hz, whilst vibration effects are primarily seen between 0.5 and 80 Hz. Over longer periods (3–4 days), adaptation to motion usually occurs and the incidence of sickness declines.

2) **Motion-induced task interruptions** – Caused by low-frequency, high-amplitude motion and specific short-term events. Abrupt changes in acceleration are frequently present, e.g. in slamming, jolting or turbulence and can have a large impact on the ability to make control inputs and read displayed information. Touch screens are not recommended for platforms susceptible to motion, especially if small and heavily influenced by sea state and touch screens are not recommended for critical control actions on any HCI.

3) **Motion-induced fatigue** – caused by low-frequency, large-amplitude motion and resulting from long-term exposure. Fatigue is due to the need for having to continually compensate for whole body displacement and to preserve suitable postures for task and other activities. Recent research has focused on better seating for small boats (e.g. ribs used by the SBS and Marines) and involves the design of the seat and where it is located on the platform.

4) **Vibration** – medium to high-frequency, with exposure time depending on tolerance to the motion severity. Vibration can severely reduce the readability of displays if not damped out.

7

Training

7.1 Outline

Training is all about the acquisition of skills. The purpose of training is to contribute to the cost-effective delivery of the required level of control room capability by preparing personnel to be at pre-defined and agreed levels of performance; optimum performance is required in the areas of physical capability, cognitive skills, knowledge and behaviour. Training also builds cohesion and teamwork and ranges from individual proficiency training to the conduct of large exercises that test command and control and the application of doctrine in a large organisation, be it a warship or an oil refinery.

Training enables personnel to continue to operate in the confusion and stress of emergency situations. Training must be provided for individual operators, users and maintainers, for control room sub-teams and whole plant teams in order to support the delivery of the system capability. Collective (Team) training plays a critical role in enabling the meeting of objectives for increased control room effectiveness with fewer personnel.

Note that well designed equipment should require minimal training. Complex cars and phones can be mastered in hours so be suspicious of procuring equipment with a high training overhead. Novice pilots can be trained to fly and land a light aircraft in 10 hours and become fully qualified to carry passengers in 50 hours! A high training overhead could mean that the equipment designers have built an unnecessarily 'difficult to use' piece of kit.

Human Factors in Control Room Design: A Practical Guide for Project Managers and Senior Engineers, First Edition. Tex Crampin.
© 2017 John Wiley & Sons Ltd. Published 2017 by John Wiley & Sons Ltd.

7.2   Training Needs Analysis and Specification

For a complex control room, training can be divided into:

1) **Individual training** – for specific isolated equipment operation;
2) **Sub-team training** – for more complex equipments requiring supervisors and operators;
3) **Full team training** – in order to hone the skills of a complete team for complex control room operations, rather like a whole orchestra.

The approach to training is to first identify the training needs, break down those needs into operational skills, then identify the best way of imparting those skills to new operators and sustain those skills for individuals, sub-teams and the full control room team.

Thus a task analysis is essential as a starting point and, for a well-designed control room, this should be already available from data collected when it was designed. Typically, a task analysis will identify an operator's role, rank or job title, career training undertaken to fulfil that Role and the specific tasks required of the role.

Typically, there are four key skills of an operator:

1) **Perceptual** – the ability to detect (through the human senses, eg, sight and hearing) control room changes and act upon them swiftly and accurately. For example, detecting an alarm embedded within a series of alarms or noticing a deviant meter reading.
2) **Procedural** – the ability to follow agreed and documented Standard Operating Procedures (SOPs) and Emergency Operating Procedures (EOPs). Training reinforces these by practice and repetition so that they become almost second nature, especially important during rapidly changing and stressful scenarios.
3) **Psychomotor** – the ability to execute demanding and precise physical tasks requiring, for example, accomplished hand-eye coordination, joystick control, etc.
4) **Cognitive** – the ability to recognise, assess and decide on a course of action; thus a thinking skill.

7.3   Training Equipment

Only when the training needs are defined can training equipment be specified. This could range from simple whiteboard classroom instruction, through part task trainers to full mission simulators. The key word is fidelity, that is, the realism required to train to the level of skill required to do the job. Some skills can be trained on very low fidelity devices whilst other skills require close to real world fidelity, achieved in an expensive simulator. It follows that a thorough Training Needs Analysis (TNA) can pay dividends in ensuring that the training budget for a control room is focused on achieving training effectiveness.

Of the four key skills introduced above, very generally, they tend to demand different levels of training equipment fidelity in order to impart the required level of training transfer and it is here that procurement costs can be reduced. Examples are provided from actual simulated training devices:

1) **Perceptual** – medium levels of fidelity because, e.g. simulated visual cues need to be realistic enough to invoke the required reaction but no more;
2) **Procedural** – low levels of fidelity because all that is required is enough simulation fidelity to invoke the correct SOPs and EOPs in the correct order;
3) **Psychomotor** – high levels of fidelity because trainees need to learn very precise sensor-muscle coordination, e.g. in manoeuvring a re-fuelling aircraft into the re-fuelling 'window' behind a donor aircraft. Too fast and the recipient aircraft may fly into the donor aircraft; too slow and the nozzle may not engage with the basket. Manual intervention of an airliner in auto-pilot requires training in very precise manipulation;
4) **Cognitive** – medium levels of fidelity because the reality of the actual equipment is not paramount in invoking a correct decision or correct course of action following a sequence of events. The scenario (mission events) that elicits the correct thinking and decision making is important.

7.4   Summary Approach to Training

In summary, the TNA process is represented diagrammatically below. Designers must be careful not to over-analyse training needs since this can take up valuable resources which would otherwise be better deployed in designing easier to use equipment that requires less training time (Figure 7.1):

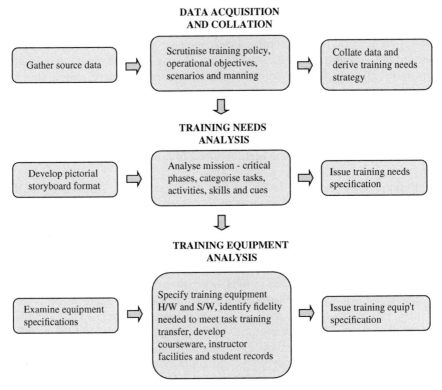

Figure 7.1 Training needs analysis (TNA) process. *Source*: Liveware library diagram copyright and publishing permission courtesy of Liveware HF Ltd.

8

Assessment and Acceptance Testing

8.1 Method

Aim: To outline practical acceptance testing for operational control rooms. The method for delivering acceptance testing covers:

1) **Acceptance Testing** – describes the reasons for and importance of acceptance testing in control rooms;
2) **HF Design Process** – identifies a practical and comprehensive HF control room design process based on proven experience with actual programmes and shows where, in this design process, assessment and acceptance is relevant;
3) **Acceptance Testing Alignment to HF** – shows where acceptance testing starts and ends within the HF design process and where the key activities reside within the process;
4) **Acceptance Testing Process** – describes the acceptance testing process in detail based on HF experience in:
   a) **Static Acceptance** – testing control room space, workplace layout, HCI format and presentation of colour, symbols, active areas, etc.;
   b) **Dynamic Scenario-based Acceptance** – testing actual operational usability with experienced control room personnel by running through realistic scenarios.

8.2 Acceptance Testing and Human Factors

Acceptance testing is a means of design validation but focusing on the usability of the system under procurement (or retrofit), be that a small equipment or major control room. Acceptance testing in this sense varies from traditional supplier equipment testing in that the user is an integral

Human Factors in Control Room Design: A Practical Guide for Project Managers and Senior Engineers, First Edition. Tex Crampin.

part of that testing process as a human-machine system. Further, all aspects are under scrutiny including:

1) The compartment Concept of Operations (CONOPS);
2) Lighting, heating and ventilation;
3) Traffic flow under all conditions;
4) Functional grouping of teams, workstation configuration and layout;
5) Detailed workstation design (HMI) including work areas, displays and controls;
6) Screen design and format (HCI) including dialogue;
7) Scenarios and operational procedures, including training.

8.3   Control Room HF Design Process and Acceptance Planning

Experience with the Royal Navy's QEC (Queen Elizabeth Class) aircraft carriers has enabled a robust and practical approach to be derived for acceptance testing. This involved quite complex human-machine systems, notably, the Operations Room, Bridge, Ship Control Centre, Flying Control, Damage Control and Fire Fighting. A summary of that process is provided below in a level of detail appropriate to being able to align it to the activities in acceptance testing. The points at where acceptance testing is relevant are shown by the grey arrows (Figure 8.1).

8.4   Acceptance Testing Detail

This method shows how simple ergonomics checklists and scenario-based walkthroughs can be used to assess control room designs early on in development and as part of the acceptance process. The two main types of HF assessment and acceptance methods are:

1) Static assessments – checklist approach;
2) Dynamic assessments – scenario based approach.

The following sub-sections focus on applying these methods, by example, to a Ship Control Centre PMS (Platform Management System), covering workspace, panel layout, display and control ergonomics. A PMS is used as an example for conveying the Assessment and Acceptance method but the assessment process applies to any complex control room system.

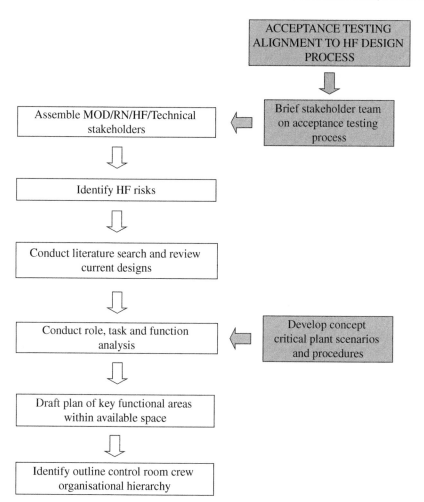

Figure 8.1 HF design process aligned to assessment and acceptance. *Source*: Reproduced with permission of Liveware HF Ltd.

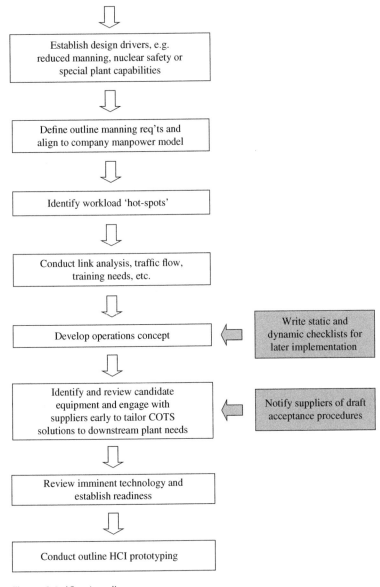

Figure 8.1 (*Continued*)

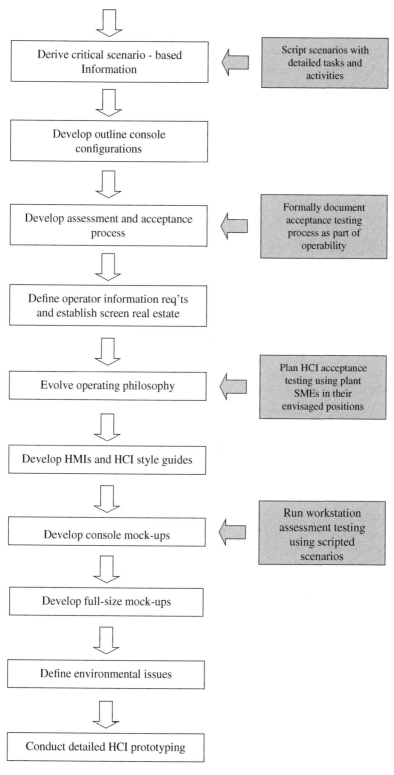

Figure 8.1 (*Continued*)

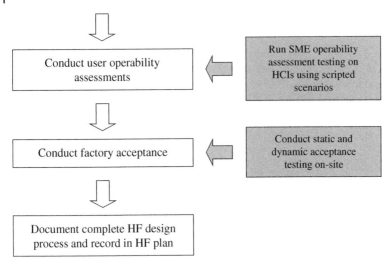

Figure 8.1 (*Continued*)

8.4.1 Static Assessments

The following static assessment procedures should be applied to workspaces, panel layouts, displays and controls. Refer to Chapter 3 for specific dimensions:

1) **Workspace** – To assess that there is sufficient space for seating access/egress and for personnel to move around the control room to facilitate smooth traffic flow as previously depicted in Figure 3.7 above.
2) **Panel Layout** – To assess that controls can be reached easily and are generally placed below displays and that displays are easily readable:
   a) Controls used in sequence should be mounted in a row or column;
   b) Controls frequently used or important for safety should be centrally mounted;
   c) Frequently used display information should be within a 30° cone around the operator's Field of View (FoV) which is 15° below the horizontal;
   d) Critical alerting information should be within a 15° cone around the operator's FoV which is 15° below the horizontal.

Static assessments involve reviewing page format designs for completeness and conformance to ergonomics principles. This provides an opportunity for early feedback before any costly programming has been applied to the screen HCI. This should cover design issues using a checklist for:

3) **Display Components** – Provision of appropriate components, that is, objects for visual reference and plant control;

4) **Colour Coding** – Adherence to accepted system standards and human stereotypes, e.g. red for danger;

5) **Consistency** – Repeated conformity to operational processes, notably those pertinent to display dialogue, e.g. starting pumps, calling up sub-pages, accepting alerts etc. should all operate in the same way so that the user is not surprised by a new process or coding dimension;

6) **Page Orientation** – Use of either a landscape or portrait format to present the most appropriate screen area for conveying the information;

7) **Labelling** – Adherence to a consistent labelling hierarchy with labels for major pages and equipment in large text, gradually reducing in size to minor components, but always readable at the design viewing distance. Generally, labelling should be above the object for large items, such as major page titles and major sub-titles. Control buttons should have the labelling placed directly on the control so that there is no doubt what the control refers to. Hovering a pointer over a control should invoke a visual change in the appearance of the control to show that it is selectable, e.g. by a raised, grey shadow.

It is important that the pages assessed can be viewed 'as will be seen' on a VDU on the final system, including any status bars and formatting that will be permanently seen.

The static assessment checklist can be re-used as an Acceptance Checklist by the Acceptance Authority.

8.4.2 Dynamic Assessments

The ergonomics checklist will identify adherence to the many basic ergonomics principles. The essence of acceptance testing is to ensure that the HCI pages will satisfy effective and efficient task completion under real and actual operational conditions. Without the opportunity to run elaborate prototype trials at an early stage, an alternative is to use paper walkthroughs of scenarios. This will help to identify how the pages work together and whether the information is easily accessible and can be provided simultaneously when required. In addition, paper walkthroughs can help to identify whether the pages are acceptable in terms of completeness.

It is suggested that only the parts of a scenario that are critical are chosen for analysis since this can be a time-consuming process. A representative sample of critical activities should be selected based upon a risk analysis and scripts detailing the scenario events and operator activities will need to be written. An example script is provided below (Table 8.1):

Table 8.1 Example outline action damage mission critical task script.

Steps	Events	PMS displayed information	Personnel and Activities	Relevant PMS Page/Control	Information shortfalls and comments	Results
	Aircraft detected port side		Officer makes Main Broadcast 'brace, brace, brace'			
A		PMS records crash stop vent	Operator crash stops vent and all personnel 'brace'	Vent Overview Zones 1–5 + Machinery Space	Check Vent HMI shows vents stopped, assess time taken to monitor these pages, investigate use of sub-pages to assist this process, assess overload of event messages	
	Aircraft attack					
B	Damage caused in Machinery Space	Auxiliary Diesel Generator No.1 trips	Operator accepts alert	Propulsion over view and Aux. DG	Check stateboard is automatically updated with availability of Aux DG	
C	Damage caused in Machinery Space	Flood alarm	Operator accepts and confirms alarm	Detailed Plant Layout Plan	Check flood appears on Plan	

D	Damage caused in Galley	Fire detected in Galley	Operator accepts and confirms alarm and calls up area description details	Detailed Plant Plan	Check aide memoire, etc, on all relevant consoles
E	Port shaft No5 plummer block overheat due to damaged shaft	Plummer block overheat	Operator accepts alert	Propulsion Overview	Check Propulsion Overview and alerts
F			Operator orders Emergency Slow port shaft	Prop Overview or Port shaft	
G		PMS records change in SIC (Station-In-Control)	Additional operator takes thrust control	Prop. Overview	Check SIC transfers
H	Repeater needle on Bridge moves into alignment		Operator reduces port shaft with lever	Port lever down	Check Propulsion Overview
I	Telegraph mismatch warning		Operator obeys telegraph on unaffected shaft		

Source: Reproduced with permission of Liveware HF Ltd.

The 'Personnel and Activities' column of the script identifies who performs which actions. This forms the basis of conducting the paper walkthrough using hardcopy VDU pages. Multiple roles may be played by an individual although the more personnel involved, the easier the process will be. Whilst the script is unlikely to be worked through in real time, SMEs (Subject Matter Experts) will be able to use their experience to identify the potential problem areas (e.g. time to access certain information, possible causes of error, etc.).

The scripts from the paper walkthroughs can be used later as part of the acceptance of the system by the Acceptance Authority. At this point performance data may be added for critical task activities, for example, how long it takes to react to an alert.

8.4.2.1 Detailed Dynamic Assessment Example

A more detailed dynamic assessment example is provided below from the author's direct experience in Royal Navy warship control room design. The Supplier should select suitable scenarios to test their HCI design solutions in collaboration with users and their individual and team tasks. The example scenario below has been selected in order to test prototyping development and evolve a CONOPS (Concept of Operation) for a ship propulsion system. The actual example is arbitrary; it is the method that is being conveyed that is the purpose of this control room guidance document.

Fire in Main Machinery Space - scenario outline:

1) **Power and Propulsion Configuration** – Port or Starboard DG running and Port and Starboard GTs and EMs running;

2) **Ship State** – Ship is on a State 3 Transit in open waters in the evening when a Fire breaks out on a DG in the Main Machinery Space (MMS). Command Aim: 'Maintain full passage speed, meet ETA at next port;'

3) **Scenario Initiation Event** – Alert on Port DG1 followed by fire detected in MMS;

4) **Operator Reactions** – The Supervisor accepts alerts in the main control room, carries out the required actions (crash stop vents, makes Main Broadcast pipe), Supervisor 2 transmits 'Emergency Slow' to bridge, starts a standby DG and stops all non-essential machinery in the affected compartment.

Scenario Script – Fire in main machinery space (Table 8.2):

Table 8.2 **Scenario script.**

Ref	Events/Activities	Comments/Questions/Issues	System Page
1	Port MMS: A fire danger occurs on the non-drive end of Port1 DG alternator. Fire danger is as a result of the lubricating cross connection pipe failing and spraying oil onto a hot exhaust.		
2	System displays Port DG1 lub oil Alert		
3	SCC: MEOOW1 - accepts Port DG1 lub oil Alert		
4	SCC: MEOOW2 - transmits Emergency Slow to Bridge	Via PDLs or telegraphs or buttons next to PDLs – Project to investigate as required	
5	SCC: MEOOW2 - selects SCC Control of Propulsion using hard wired buttons next to PDLs		
6	SCC: MEOOW2 - starts Stbd DG		
7	SCC: MEOOW1 - briefs OOW via conning	Needs conning available at his workstation	
8	SCC: MEOOW1 - sends out roundsman to investigate		
9	Port MMS: Port1 DG develops a serious fire which is being fed by the engine oil supply. Flame detector above Port1 DG detects the fire and sounds the alert via the Minerva detection System in HQ1	Fire Alert annunciates on dedicated Fire Alerts panels on bridge and in HQ1 and on system. Lub Oil Alert annunciates on system.	
10	SCC: MEOOW1 - accepts Fire alert on Minerva panel		Fire Detection Mimic
11	SCC: MEOOW1 - crash stops vent using hard wired crash stop on MEOOW1's console.	DCO will instruct MEOOW1 to crash stop vents if in compartment,	
12	SCC: MEOOW1 - makes Main Broadcast 'Fire, Fire, Fire...........	Will MEOOW1 have Main Broadcast comms at his console?	

(*Continued*)

Table 8.2 (Continued)

Ref	Events/Activities	Comments/Questions/Issues	System Page
13	SCC: MEOOW1 - trips Port DG1 using hard wired trip once standby DG is on load.	DG and GT trips need to be provided at MEOOW1's workstations (in between MEOOW1 and MEOOW2's workstations?) Tripping DG will get it shut down faster than a controlled stop (via the system Display pages)	
14	SCC: MEOOW1 - reports to OOW on Conning 'Ready to obey telegraphs' and passes Propulsion Limitations to Bridge	OOW may stay in SCC control or revert to Bridge control of propulsion.	

Source: Reproduced with permission of Liveware HF Ltd.

References

1 Ministry of Defence. Human Factors for Designers of Systems; 2008. Defence Standard 00-250: Issue 1, published 23 May 2008. Def Stan 00-250 is managed by TES-DStan-SPM4 on behalf of the Ministry of Defence. Any enquiries on this standard should initially be directed to UK Defence Standardization (DStan): Standards Programme Manager, UK Defence Standardization, Room 1138, Kentigern House, 65 Brown Street, Glasgow G2 8EX.

2 Ministry of Defence. Requirements for Damage Surveillance and Control Management Systems in HM Surface Ships, Submarines and Royal Fleet Auxiliaries; 2014. Defence Standard 08-111, published 24 July 2014. MOD, Abbey Wood, Bristol, BS34 8JH.

3 Department of Defense. Human Engineering; 2012. Design Criteria Standard: MIL-STD-1472G. Program Office Wright-Patterson Air Force Base, OH USA 45433-657.

4 Health and Safety Executive. Human Factors and Ergonomics; 1999. HSG48. http://www.hse.gov.uk/humanfactors

5 Carr, N. The Glass Cage. 1st edn. London: The Bodley Head; 2015.

6 Lee, JD. Human Factors and Ergonomics in Automation Design. In Handbook of Human Factors and Ergonomics, ed. Gavriel Salvendy. 3rd edn. Hoboken, NJ: John Wiley & Sons, Inc.; 2006.

7 Ministry of Defence Human Factors Integration for Defence Systems; 2013. JSP 912: Version 2i (internet) 25th June 2013. Sponsored by Director Technical (DTech), Defence Equipment and Support (DE and S). Produced and maintained by the Engineering Group (EG), DE and S. Issued under the authority of DE and S, a branch of the MOD and available online through MOD JSP.

Human Factors in Control Room Design: A Practical Guide for Project Managers and Senior Engineers, First Edition. Tex Crampin.
© 2017 John Wiley & Sons Ltd. Published 2017 by John Wiley & Sons Ltd.

Index

Note: Page numbers appearing in bold indicate tables, page numbers in *italic* indicate figures.

Human Factors in Control Room Design: A Practical Guide for Project Managers and Senior Engineers, First Edition. Tex Crampin.
© 2017 John Wiley & Sons Ltd. Published 2017 by John Wiley & Sons Ltd.

Printed and bound by CPI Group (UK) Ltd, Croydon, CR0 4YY

06/05/2024

14497892-0001